Analytical Chemistry

REFRESHER MANUAL

T0200575

John Kenkel

CRC Press
Taylor & Francis Group
Boca Raton London New York

CRC Press is an imprint of the
Taylor & Francis Group, an **informa** business

Published 1992 by CRC Press LLC
Taylor & Francis Group
6000 Broken Sound Parkway NW, Suite 300
Boca Raton, FL 33487-2742

©1992 by Taylor & Francis Group, LLC
CRC Press is an imprint of Taylor & Francis Group, an Informa business

First issued in paperback 2019

No claim to original U.S. Government works

ISBN-13: 978-0-367-45037-3 (pbk)
ISBN-13: 978-0-87371-398-6 (hbk)

Visit the Taylor & Francis Web site at
http://www.taylorandfrancis.com

and the CRC Press Web site at
http://www.crcpress.com

Library of Congress Cataloging-in-Publication Data

Kenkel, John.
Analytical chemistry refresher manual/by John Kenkel.
    p. cm.
  Includes and index.
    ISBN 0-87371-398-2
1. Chemistry, Analytic.   I. Title.
QD75.2.K447   1992
543—dc20   91-38547

Library of Congress Card Number 91-38547

DEDICATION

This book is dedicated to my mother, Carmen Kenkel, who, following the death of my father when I was quite young, functioned as both mother and father for me during the years when I was growing up. In spite of her sudden death in 1974, she continues to this day to be a major source of inspiration for me and there is no doubt that her prayers on my behalf from her present heavenly home have resulted in the divine intervention needed to begin and complete this book. Thanks, Mom.

DEDICATION

This book is dedicated to my mother, Carmen Kenkel, who, following the death of my father when I was quite young, functioned as both mother and father for me during the years when I was growing up. In spite of her sudden death in 1974, she continues to this day to be a major source of inspiration for me and there is no doubt that her prayers on my behalf from her present heavenly home have resulted in the divine intervention needed to begin and complete this book. Thanks, Mom.

AUTHOR

John Kenkel has been on the chemistry faculty at Southeast Community College in Lincoln, Nebraska, since 1977. He has been chair of the Environmental Laboratory Technology training program there since 1979. In this position, he has trained more than 300 students in the techniques of modern analytical chemistry, including both wet methods and instrumental methods. Graduates of his program are acclaimed as outstanding people for technician-level laboratory positions.

Professor Kenkel, born and raised on a small farm in western Iowa, has a B.A. from Iowa State University (1970) and a M.A. in chemistry from the University of Texas at Austin (1972). While at Texas, he worked under Professor Allen Bard in electroanalytical chemistry. He was employed as a corrosion chemist at the Science Center, Rockwell International, Thousand Oaks, California, from 1973 to 1977. He has co-authored more than 15 articles stemming from his research at the University of Texas and at Rockwell and also 2 papers relating to chemistry technician training.

His first book, "Analytical Chemistry For Technicians", published by Lewis Publishers/CRC Press in 1988, has become a primary training manual for analytical chemistry in 2-year college laboratory technician training programs across the country.

Professor Kenkel is a long-standing member of the 2-year college chemistry organization, $2YC_3$, and of the $2YC_3$ Advisory Board, midwest region. He is also a member of the American Chemical Society and has been active in the ACS Division of Analytical Chemistry as a member of their Committee on Education. He is the chairman of the subcommittee on Associate Degree Education. As a member of this subcommittee, he has created and is editor of the Newsletter for Chemistry Technician Instructors (NCTI), a semiannual newsletter for other instructors in this field.

He has won several awards for teaching excellence. In 1985, he was awarded the Burlington Northern Faculty Achievement Award at Southeast Community College. In 1988, he won a regional Catalyst award for excellence in college chemistry teaching. This award is sponsored by the Chemical Manufacturers Association. In 1989, he won the Ohaus/National Science Teachers Association award for innovation in college science teaching. He has twice been a nominee for the Wekesser Outstanding Teaching Award at Southeast Community College.

PREFACE

This work is intended to be, as the title implies, a refresher in the techniques and methodology of modern analytical chemistry. It is directed specifically toward technicians and other workers in the analytical chemistry laboratory who find that they need a personal reference manual to help acquaint (or reacquaint) them with traditional and nontraditional chemical analysis techniques that may have unexpectedly become part of their jobs. These individuals may have received no formal training in this discipline or may have received formal training but, given the rapid growth of this science in recent years, find themselves sinking in the maze of procedures and special techniques that are encountered today.

It is intended to be a book for individuals who were biology majors, geology majors, environmental science majors, etc., who find that they are expected to function as chemical analysis technicians. It is intended to be a fundamental, readable reference on analytical chemistry packed full of tidbits of theory and background highlights of all popular analytical methods.

The book includes information on sampling and sample preparation, solution preparation, and discussions of wet and instrumental methods of analysis. The spectrometric techniques of UV, vis, and IR spectroscopy, including FTIR, are fully covered. NMR and mass spectrometry, as well as the full complement of atomic spectroscopy techniques such as flame photometry, flame AA, graphite furnace

AA, and ICP, are covered. A thorough discussion of analytical separations, including liquid-liquid extraction, liquid-solid extraction, both instrumental and noninstrumental chromatography, and electrophoresis, are covered. Gas chromatography and high performance liquid chromatography are given separate chapters in which the basic theory and instrument design concepts are covered. Potentiometric and voltammetric techniques are presented followed by a chapter on automation. The final chapter in the book discusses the detection and accounting of laboratory errors.

The author hopes that laboratory workers at all levels in all types of applications, from environmental analysis to quality assurance, from private testing laboratories to government analysis laboratories, from the manufacturing industry to water and wastewater treatment, and from routine work to academic research, will benefit.

ACKNOWLEDGMENTS

The *Analytical Chemistry Refresher Manual* would not have happened were it not for the help and encouragement of a number of people. I would first like to thank members of the analytical chemistry community for supporting the concept of analytical chemistry books for the technician-level laboratory worker, and I hope that this book will serve their needs.

Second, I would like to thank my editor, Brian Lewis of Lewis Publishers, Inc., for suggesting this book and for his patience and encouragement as the project unfolded. It has been a pleasure to work with him.

Third, I say "thank you" to those chemistry professionals whose names are unknown who read my initial outline and made suggestions. Their help in developing the book in its final form was very important.

Several people read the first draft of the manuscript and offered suggestions. These include Brad Godwin of PACE Laboratories, Inc. of Lenexa, Kansas, and Reza Rafat of SmithKline Beecham Animal Health in Lincoln, Nebraska, and another who remains anonymous. Thank you all for your very positive contributions.

I would like to thank my primary artist, Lana Johnson of Lincoln, Nebraska, who drew most of the illustrations in the book. She has had a very major impact in the book's quality, and I appreciate it very much.

I sincerely appreciate the love and understanding of my wife, Lois, and children, Angie, Jean, and Laura, throughout this period. Without their smiling faces and their understanding, completing this book would have been a much greater burden. As it was, it was a very pleasant experience.

Finally, and most important of all, I want to acknowledge my heavenly Father for His many blessings on me and my family, especially for the gifts of life and happiness that come from Him. This book would only have been a passing thought if not for Him.

John Kenkel
Southeast Community College
Lincoln, Nebraska

TABLE OF CONTENTS

CHAPTER 1

INTRODUCTION TO ANALYTICAL CHEMISTRY

1.1 IMPORTANCE OF ANALYTICAL CHEMISTRY

We begin this refresher on analytical chemistry with the following quotation from an editorial which appeared in a recent issue of *Chemical and Engineering News*, a publication of the American Chemical Society, Washington, D.C.

> "This should be a golden age for chemists. And in many ways it is. An unending progression of new instruments and computers continues to accelerate the pace of chemical research. Determinations that once took years and were the stuff of doctoral theses can now be done routinely overnight. Applications of chemistry to interdisciplinary areas are multiplying both the scope of chemical research and the range of the chemists' contributions to the public good. The intellectual contributions of chemists to the understanding of nature remain exciting, rewarding, and boundless. The industries that chemists have created remain basically sound and progressive. The role of chemists in handling health and environmental problems is as vital as ever and growing."[*]

While Mr. Heylin's statement precedes an editorial on the critical

---

[*]   Excerpted from *Chem. Eng. News*, 68(5), 3, 1990. With permission. Copyright 1990 American Chemical Society.

state of chemistry as a profession in today's world, it could just as easily precede an editorial on the current and future challenges that face analytical chemists today. It is truly a golden age. Instrument and computer manufacturers are producing analytical machines that are ever-increasing in power and scope. Sophisticated analytical determinations are indeed becoming routine. Analytical chemistry now touches more interdisciplinary areas and plays a greater role in the multiplication of the scope of the chemical science in general than ever before. The intellectual contributions of analytical chemists to the understanding of nature, to the chemical industries, and to the sciences of health and the environment are, to be sure, exciting, rewarding, sound, progressive, vital, and growing.

The most intriguing adjective Mr. Heylin uses, however, is "boundless." Such an adjective in effect says that there is no limit to the scope of the discipline. This word describes modern analytical chemistry perhaps better than any other. Today, we expect analytical chemists to be a vital link in an extraordinary number of diverse fields. Industries that manufacture chemicals, pharmaceuticals, food products, paints, polymers, plastics, indeed almost any consumable product we can identify, utilize analytical chemists for quality assurance and for the research and development of new products. Environmental and health scientists today rely heavily on analytical chemists for rapid and accurate analysis of air, water, food, soil, plants, waste, and biological samples for potentially harmful chemicals. Agricultural scientists, to a large extent, rely on analytical chemists for the analysis of feeds, fertilizers, water, soil, plants, and biological samples for chemical additives and residues. Medical doctors, and thus the general public, have come to rely on analytical chemistry for an accurate analysis of blood, urine, skin, and other biological tissue and fluids. Analytical chemistry has found its way into such fields as veterinary science, archaeology, farming, and space travel. The recent controversies surrounding the Shroud of Turin (purported burial cloth of Jesus Christ) and its analysis indicates that even religion is within its scope. The modern principles and applications of analytical chemistry in today's world are as important and as relevant to our existence as any other field that exists. Figure 1.1 presents an illustration of the dependence of various professional individuals and groups on the services that an analytical chemist renders.

The purpose of this text is to provide readable and timely descriptions of fundamental analytical laboratory processes starting with the very

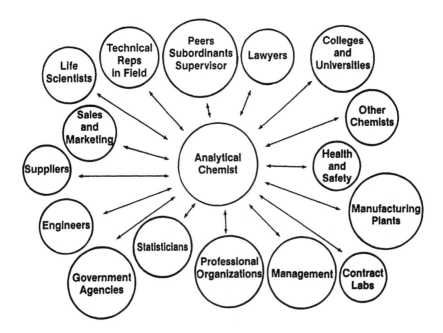

FIGURE 1.1   Illustration of numerous interactions that an analytical chemist can have with other professional individuals and groups in the modern world. (Courtesy of Steven Zumdahl, University of Illinois, Champaign-Urbana, IL.)

basic premises upon which all chemical analyses depend, that of the precision and accuracy of measurement, the fundamentals of sampling, and the thought processes involved in the planning and execution of the actual lab work itself.

1.2 PRECISION VS ACCURACY

Analytical chemistry is the science dealing with the determination of the chemical makeup of real-world material samples, with the major objective often being the numerical expression of the quantity of a component present in that sample. The accuracy of this expression is obviously very important. Analysts should therefore always be asking themselves if there is any reason to doubt the accuracy of their results. They must give some thought as to what may be the causes of *inaccurate* results. Errors on the part of the analyst himself or of the equipment or chemicals used are obvious causes. There may be many potential sources of such error in a given experiment, some of which can be determined (determinate errors) and some of which cannot (indeterminate errors).

Determinate errors are errors which are known to have occurred. They can arise from contamination, wrongly calibrated instruments, carelessness, etc. They can be avoided by exercising careful laboratory practices and technique. If it is determined that such an error occurred, then one obviously cannot trust the answer to be accurate.

Indeterminate errors, however, are impossible to avoid. They are "random" errors; errors which either were not known to have occurred or were known to have occurred, but could neither be taken into account, nor avoided. They are often errors which occur each time a particular analysis is run. Such errors can determine the predicted quality of the results, such as the sensitivity and detection limits of a given experiment. The error in reading a meniscus and the error associated with an instrument readout are examples of this type of error. These errors must be dealt with by statistics and the laws of probability. One way is to repeat a given analysis or procedure a number of times and use statistics to determine whether the results are precise. The determination of a "confidence interval" is often the result of a statistical analysis. (For example, the concentration of nitrate in a water sample may be expressed as 5.4± 0.2 ppm. The confidence interval is thus 5.2–5.6 ppm.) We consider precise data as having a high probability of being accurate.

The two terms "precision" and "accuracy" are often misrepresented. Accuracy refers to the "correctness" of a given measurement or result. That is, does the measurement (or measurements) come close to what the correct answer is or what it is expected to be? A "control" sample is often run alongside the unknowns as a check on accuracy. Precision does not necessarily relate to accuracy. Precision refers to how well a series of identical measurements on the same sample agree with each other. Figure 1.2 represents a classical illustration of the difference between accuracy and precision. While precision is not necessarily synonymous with accuracy, it is often taken to indicate accuracy, unless a control indicates otherwise or unless there is known to be a factor which inherently affects accuracy on a continuous and constant basis. Chapter 13 presents more information on data quality and methods of handling laboratory data.

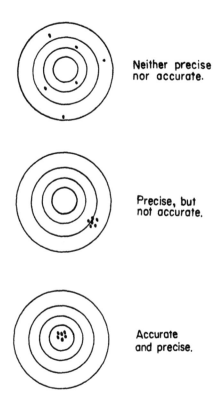

FIGURE 1.2   Illustration of precision and accuracy. (Reprinted from Kenkel, J., *Analytical Chemistry for Technicians*, Lewis Publishers, Chelsea, MI, 1988. With permission.)

1.3 TERMINOLOGY

The basic terminology associated with analytical chemistry and analytical laboratory work is important and may be foreign to persons who have not been associated with such laboratory work in preparation for their jobs. We thus present a small glossary of terms in this section. Other important terms specific to particular analyses are given elsewhere in this book and can be found in the index.

Chemical Analysis   This is the determination of the chemical composition or chemical makeup of a material sample.

Qualitative Analysis   The determination of *what* substances are present in a material sample, usually without

	the need or desire to determine quantity of these substances.
Quantitative Analysis	The determination of *how much* of a specified substance is present in a material sample.
Quantitation	This is the determination of quantity, as in the quantitative analysis above.
Quantification	This is another word for quantitation.
Quantitative Transfer	A transfer of a chemical or solution from one container to another, making sure that every trace of this chemical is in fact transferred.
Analyte	This is the substance being analyzed for in an analytical procedure. This can be an element, a compound, or an ion.
Assay	This is another word for chemical analysis.

1.4 FUNDAMENTALS OF MEASUREMENT

In the analytical chemistry laboratory, many measurements are made, and the accuracy of these measurements obviously is a very important consideration. Different measuring devices give us different degrees of accuracy. A measurement of 0.1427 g is more accurate than a measurement of 0.14 g simply because it contains more digits. The former (0.1427 g) was made on an analytical balance, while the latter was make on an ordinary balance. A measurement recorded in a notebook should always reflect the accuracy of the measuring device. It does not make sense to use a very accurate measuring device and then record a number that is less accurate. For example, suppose a weight on an analytical balance was found to be 0.14g. It would be a mistake to record the weight as 0.14 g, even if you know personally that the weight is 0.14 g. Presumably, there are other people in the laboratory using the notebook, and your entry will be construed as to contain only two digits. The following example further illustrates this point.

Figure 1.3 shows a meter, such as a pH meter, or readout meter on the face of some other laboratory instrument. The correct reading on this meter is 56.7. The temptation may be to write down 57. This latter reading is not correct in the sense that it does not reflect the accuracy of this measuring device. Measuring devices should always be used

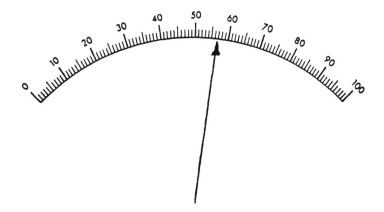

FIGURE 1.3   A measuring device registering a reading of 56.7 and not 57. (Reprinted from Kenkel, J., *Analytical Chemistry for Technicians*, Lewis Publishers, Chelsea, MI, 1988. With permission.)

to their optimum capability, and this means recording all the digits that are possible from the device. The general rule of thumb for a device such as the meter in Figure 1.2 is to write down all the digits you know with certainty and then estimate one more. Obviously, the meter in Figure 1.3 shows a reading between 56 and 57, or "56 point something." This "something" is the estimated digit and is estimated to be a 7. The correct reading is 56.7. For digital readouts, such as with an analytical balance, this estimation is done for you by the device.

The digits that are actually part of analytical measurements have come to be known as "significant figures." A knowledge of the subject of significant figures is important from the standpoint that (1) one needs to know the accuracy of a measurement from just seeing it in a notebook (and not necessarily from actually seeing it displayed on the measuring device), (2) calculations are usually performed using the measurement, and (3) the correct number of significant figures must be shown in the result of the calculation.

To cover point number 1 above, the following rules apply.

1.   Any nonzero digit is significant. Example, 1.27 (three significant figures.)
2.   Any zero located between nonzero digits is significant regardless of the position of the decimal point. Example, 1.027 (four significant figures).
3.   Any zero to the left of nonzero digits is not significant, unless it

is covered also by Rule #2. Such zeros are shown only to locate the decimal point. Example, 0.0127 (three significant figures).

4.   Any zero to the right of nonzero digits and to the right of a decimal point is significant. Example, 1.270.

5.   Any zero to the right of nonzero digits and to the left of a decimal point may or may not be significant. Such zeros may be shown only to locate the decimal point or they may be part of the measurement; one does not know unless he/she personally made the measurement. Such numbers are actually incorrectly recorded. They should be expressed in scientific notation to show the significance of the zero because then Rule #4 would apply.

Example,

1270. (incorrect)

1.270×10^3 (four significant figures) or 1.27×10^3 (three significant figures)

To cover point number 2, the following rules apply:

1.   The correct answer to a multiplication or division calculation must have the same number of significant figures as in the number with the least significant figures used in the calculation. Example, $1.27 \times 4.6 = 5.842 = 5.8$. Hence, 5.842 is the calculator answer, and 5.8 is the correctly rounded answer.

2.   The correct answer to an addition or subtraction has the same number of digits to the right of the decimal point as in the number with the least such digits in the calculation.

Example,

$$4.271 + 6.96 = 11.231 = 11.23$$

11.231 is the calculator answer, and 11.23 is the correctly rounded answer.

3.   When several steps are required in a calculation, no rounding would take place until the final step.

4.   When both Rules #1 and #2 apply in the same calculation, the number of significant figures in the answer is determined by

following both rules in the order in which they are needed, keeping in mind that Rule #3 also applies:

Example,

$$(7.27 - 4.8) \times 56.27 = 138.9869 = 1.4 \times 10^2$$

138.9869 is the calculator answer, and 1.4×10^2 is the correctly expressed answer.

5.   In cases in which a conversion factor that is an exact number is used in a calculation, the number of significant figures in the answer depends on all other numbers used in the calculation and not this conversion factor. To say that a number is exact means that it has an infinite number of significant figures and as such would never limit the number of significant figures in the answer.

Example,

$$1.247 \text{ m} \times 100 \text{ cm/m} = 124.7 \text{ cm} \text{ (four significant figures)}$$

6.   In cases in which the logarithm of a number needs to be determined, such as in converting [H⁺] to pH or in the conversion of percent transmittance to absorbance, the number of digits in the mantissa of the logarithm (the series of digits to the right of the decimal point) must equal the number of significant figures in the original number.

Example,

$$\left[H^+\right] = 4.9 \times 10^{-6} \text{ M}$$

$$pH = -\log\left[H^+\right] = 5.31$$

This would appear to be an increase in the number of significant figures compared to the original number (three in 5.31 and only two in 4.9×10^{-6}), but the characteristic of the logarithm, the digit(s) to the left of the decimal point, represents the exponent of 10, which serves to only designate the position of the decimal point in the original number and, as such, is not significant.

1.5 SAMPLING AND PLANNING

1.5.1 Obtaining the Sample

A laboratory analysis is almost always meant to give a result which is indicative of a concentration in a very large system. A farmer wants an analysis result to represent the concentration for an entire 40-acre field. A pharmaceutical manufacturer wants an analysis result to represent the concentration of an active ingredient in 80 cases of its product, each case containing three dozen bottles of 100 tablets each. A governmental environmental control agency wants a single laboratory analysis to represent the concentration of a toxic chemical in every cubic inch of soil within 5 mi of a hazardous waste dump site.

An ideal analysis is one in which the entire system can be run through the analysis procedure. Obviously, this is not possible in most cases. The analyst is then faced with the serious problem of obtaining a sample from this system which can only be as good as the sample itself.

Obviously, there are different degrees of difficulty and different sampling modes involved with obtaining samples for analysis, depending on the type of sample to be gathered, whether the source of the sample is homogeneous, the location of and access to the systems, etc. For example, obtaining a sample of blood from a hospital patient is completely different from obtaining a sample of coal from a train car full of coal.

As far as blood is concerned, from what part of the body to sample a person's blood needs to be considered. Second, the time of day, along with a knowledge of the patient's recent diet, is important. Third, perhaps the patient is on some sort of medication which could affect the analysis.

With the coal sample, it is important to recognize that the coal held in a train car may not be homogeneous, and a sampling scheme which includes taking samples from different parts of the whole system must be implemented.

The key word in any case is "representative." A laboratory analysis sample must be representative of the whole so that the final result of the chemical analysis represents the entire system that it is intended to represent. If there are variations, or at least suspected variations, in the system, small samples must be taken from all suspect locations.

If results for the entire system are to be reported, these small samples are then mixed well to give the final sample to be tested. In some cases, analysis on the individual samples may be more appropriate. Some examples follow.

Consider the analysis of soil from a farmer's field. The farmer wants to know whether he needs to apply a nitrogen-containing fertilizer to his field. It is conceivable that different parts of the field could provide different types of samples in terms of nitrogen content, particularly if there is a cattle feed lot nearby, perhaps uphill from part of the field and downhill from another part of the field. Obviously, the sample taken should include portions from all parts of the field which may be different so that it will truly represent the entire field. Alternatively, two samples could be taken, one from above the feed lot and one from below the feed lot, so that two analyses are performed and reported to the farmer. At any rate, one wants the results of the chemical analysis to be correct for the entire area for which the analysis is intended.

Consider the analysis of the leaves on a tree for pesticide residue. The tree grower wants to know if the level of pesticide residue on the leaves indicates whether the tree needs another pesticide application. Once again, the analyst must consider all parts of the tree that might be different. Leaves at the top, in the middle, and at the bottom should be sampled (one can imagine differences in application rates at the different heights); leaves on the outside and leaves close to the trunk should be sampled; and perhaps there would also be differences between the shady side and the sunny side of the tree. All leaf samples can then be combined and brought into the laboratory as a single sample.

Consider the analysis of a blood sample for alcohol content (imagine that a police officer suspects a motorist to be intoxicated). The problem here is not sampling different locations within a system, but rather a time factor. The blood must be sampled within a particular time frame which would demonstrate intoxication at the time the motorist was stopped.

The problems associated with sampling are unique to every individual situation. The analyst simply needs to take all possible variations into account when obtaining the sample so that the sample taken to the laboratory truly does represent what it is intended to represent. A representative sample is one which has all the characteristics — all the components at their respective average concentration levels — of the whole from which it is taken.

1.5.2 Handling the Sample

How to get the sample from the sampling site to the laboratory without contamination or alteration is generally not as challenging, in most cases, as the problem of how to obtain a representative sample. There are basically two considerations associated with such sample preservation: (1) storage of the sample in a container which will protect the integrity of the sample, particularly if trace constituents are to be determined, and (2) preservation of the sample from problems which may be internal, for example, temperature effects or bacterial effects.

If trace amounts of metals are to be determined, for example, one would not want to store the sample in a glass container, since glass can leach small amounts of metals. On the other hand, if trace organics are to be determined, plastic containers may be deemed inappropriate. Sometimes, refrigeration may be important to avoid decomposition from bacterial sources. At any rate, proper sample handling methods must be employed to ensure sample integrity.

1.5.3 Planning the Lab Work

Once the sampling and sample preservation schemes have been properly performed, the sample is in the laboratory and ready for the analysis. The subsequent laboratory work will often involve dissolving the sample; preparing standards; weighing and adding sample additives, such as pH adjusters, oxidizing agents and reducing agents, color developing agents, chelating agents, and the like, before any analytical measurements are made. What tasks are actually performed and how they are performed depends on the object of the analysis.

As mentioned previously, analytical work may either be qualitative or quantitative. With qualitative analysis, we are merely interested in identifying a constituent of the sample and not determining quantity. The most important consideration here may be to be certain that the sample is not contaminated with a chemical that would interfere with this identification. For example, infrared spectrometry is often used as a qualitative tool. As will be discussed in Chapter 6, contaminated samples used in this technique yield results which may confuse the analyst. On the other hand, a dilution, or other change in the concentration of an analyte in a sample, will not generally affect a qualitative analysis, as long as the analyte remains above the detection limit of the technique chosen. Thus, when planning a qualitative analysis, the use of precise measuring devices, such as analytical balances and volumetric pipettes, may be excluded from the plan.

A quantitative analytical procedure, however, will certainly involve measurements and methods which need to be as accurate as possible, but will also involve measurements and methods which need not be accurate at all. *An efficient analytical technician is one who is able to tell the difference.* What follows is a typical cookbook analytical procedure, the spectrophotometric determination of manganese in steel, and a discussion of this procedure, step by step, to illustrate the reasoning required to discover when accuracy is important and when it is not. As a general guideline, we can say that accuracy is important when the lack of accuracy may adversely affect either the quantity of analyte ultimately measured or the other parameters which are measured and entered into the calculation of the results.

1.5.3a Example Analysis — The Spectrophotometric Determination of Manganese in Steel

Step 1.   Calculate the volumes of 1000 ppm Mn required to prepare standard solutions of 2, 5, 10, 15, and 20 ppm Mn to be placed ultimately in 100-mL flasks. Measure these volumes into separate 250-mL beakers. Add 25 mL of dilute nitric acid (25% by volume) to each. Use a sixth beaker, containing 25 mL of the nitric acid, for a blank. Cover with watch glasses.

Step 2.   Weigh a sample of the steel (0.25 to 0.30 g) into a seventh beaker and also add the 25 mL of the dilute nitric acid. Cover this beaker with a watch glass and simmer on a hot plate in a fume hood until the steel is dissolved or until only a small amount of carbon remains.

Step 3.   Remove from the hot plate and allow to cool for 10 min. Then add 0.50 g of ammonium peroxydisulfate to each of the seven beakers and bring all seven solutions, with watch glasses in place, to boil in a fume hood and boil gently for 10–15 min. The purpose of this step is to oxidize the carbon compounds in the steel.

Step 4.   The color developing agent, the oxidizing agent used to quantitatively oxidize the manganese to permanganate at this point, is KIO_4. Add, to each beaker, 25 mL of water, 5 mL of concentrated H_3PO_4, and 0.4 g of KIO_4. Return each to the hot plate and boil again for 5 min (watch glasses in place). Cool again and transfer quantitatively to 100-mL volumetric flasks. Dilute all flasks to the mark and shake.

Step 5.   Measure the absorbance of the six solutions (standards and sample), using the blank for calibration, at 522 nm. Plot absorbance vs concentration and obtain the concentration of the sample solution. Calculate the ppm Mn in the steel as follows:

$$\text{ppm Mn} = \frac{C_{\text{sample}} \times 0.100}{\text{kg of steel used}}$$

1.5.3b Discussion of Example Analysis

1.   The measurement of the volumes of 1000 ppm Mn placed in the beakers is ultimately the basis of the standard curve plotted in Step 5. These volumes will directly affect determination of the concentration of the manganese in the sample solution, and therefore they need to be accurate. The nitric acid, however, is needed to approximate the matrix of the sample solution, since nitric acid is needed for the dissolution (Step 2). It will not directly affect the quantity of manganese in the solutions and thus does not need to be accurately measured.

2.   The weight of the steel needs to be as accurate as possible, since it enters directly into the calculation of the results in Step 5. Watch glasses are important to prevent losing the manganese via splashing if and when the solution boils.

3.   Since the purpose of the ammonium peroxydisulfate is to oxidize the carbon compounds in the steel, it does not directly affect the quantity of Mn measured nor does it enter into the calculation in Step 5, and it does not need to be measured accurately. An excess of such chemical additives is generally of little consequence. In this case, a potential interference is eliminated.

4.   The KIO_4 in this step is like the ammonium peroxydisulfate in Step 3. It is important to get the Mn in the correct oxidation state, but some excess is unimportant. Its measurement does not have to be accurate. The same is true of the H_3PO_4. A quantitative transfer of the solutions to the volumetric flasks is essential, since the manganese concentration in the standards and sample will be

affected if it is not quantitative. The 100-mL volumes must be accurate, since again the concentration of the Mn will be affected and also because this volume enters directly into the calculation in Step 5 (the 0.100 is liters of solution).

Another important aspect of "planning" is deciding on an approach to an analysis, including the technique or techniques to use if one is not specified. The presentation of the theory and application of the myriad of wet and instrumental techniques and procedures available to the analyst is the primary objective of this text and cannot be handled here in a few paragraphs. However, we can present a summary of the application of these techniques so that you can refer to the appropriate chapter when requiring more information.

A summary of the applications of the major analytical techniques for use in this manner is given in Table 1.1. This table is *not* intended to be our only answer to the question, however. To know what to do when presented with an environmental water sample, a soil sample, a dried paint sample, a rag with something on it, or a chunk of concrete is often a major problem. Sample preparation schemes associated with such problems — schemes required to prepare a sample for one of the techniques in Table 1.1 — are also very important. Chapters 2 and 8 give additional information on this subject, including methods of complete dissolution of samples and also liquid and solid extraction.

Finally, because of the value and importance that are usually riding on the results of a chemical analysis, either qualitative or quantitative, great care must be exercised in the lab when handling the sample and all peripheral materials. Contamination and/or loss of sample through avoidable accidental means cannot be tolerated. The results of a chemical analysis could affect such ominous decisions as the freedom or incarceration of a prisoner on trial, whether to proceed with a move that could mean the loss of one million dollars for an industrial company, or the life or death of a hospital patient.

Laboratory workers must develop a kind of psychology for operating in a chemical analysis laboratory. They should always stop and think before proceeding with a new step in the procedure:. "What might happen in this step that would cause contamination or loss of the sample?" In this way, the avoidable (determinate) errors can be minimized.

Table 1.1   A Summary of the Applications of the Major Analytical Techniques

Technique	General Application
Gravimetric analysis	Analytes separated by physical means and easily weighed
Titrimetric analysis	Analytes present at high concentrations or for which there are convenient, well-established methods
UV-vis spectrophotometry	Molecular and ionic analytes capable of absorbing ultraviolet or visible wavelengths while in dilute solutions
Infrared spectrophotometry	Pure molecular analytes only
Fluorometry	Molecular or ionic analytes capable of fluorescence or fluorescence quenching while in dilute solution
Atomic absorption and emission	Metals in dilute solution, natural liquids, and extracts and solutions of solids
Gas chromatography	Mixtures of volatile organics, organic solvent extracts, and gases
High performance liquid chromatography (HPLC)	Complex mixtures (solutions) of analytes, including liquids and solids, organic and inorganic
Electroanalytical chemistry	Virtually all forms of analytes, metals, and nonmetals, molecular and ionic, capable of oxidation or reduction with small voltages at an electrode surface or capable of being detected potentiometrically

CHAPTER 2

BASIC CHEMICAL ANALYSIS TOOLS — DESCRIPTION AND USE

2.1 INTRODUCTION

Without some introduction to the general basic tools of an analytical laboratory, the novice in the laboratory may easily become confused and frustrated. "What is the difference between a top loading balance and an analytical balance?" "When do I use an Erlenmeyer flask as opposed to a volumetric flask?" "Why do you want me to dispense the solution from the buret with my hands in such an uncomfortable position?" "I thought any acid would dissolve this piece of copper!" These questions and statements are likely to be uttered by many a novice. A discussion of basic analytical laboratory tools is therefore appropriate and necessary.

2.2 BALANCES

2.2.1 The Basic Concept of the Balance

The most fundamental, and possibly the most frequent, measurement made in an analysis laboratory is that of weight (or mass). While we speak of mass and weight often in the same breath, it is of some importance to recognize that they are not the same. Mass is the "quantity" or "amount" of a substance being measured. This quantity is the same no matter where the measurement is made — on the midwestern plains, on the highest mountain peak, at sea level, or on the surface of Mars. Weight is a measure of the gravitational force exerted on a quantity of matter in a certain locale. Weight is one way to measure mass. In other words, we can measure the quantity of a substance by measuring the gravitational effect on it. Since 100% of all weight measurements made in any analysis laboratory are made on the surface of the earth where the gravitational effect is nearly constant, weight has become the normal method of measuring mass. Thus, mass and weight have come to be interchangeable quantities even to the point of the unit involved. Weighing devices are calibrated in grams, which is defined as the basic unit of mass in the metric system.

The laboratory instrument built for measuring mass is called the "balance." A balance is a device in which the weight of an object is determined by balancing the object usually across a knife-edge fulcrum with a series of known weights on the other side, as shown in Figure 2.1. Older balances used in analytical laboratories closely resembled this basic design, having a pointer in the center to indicate when a balance between the two pans was achieved. The sum of the weights on the "known" pan was then calculated, and the answer was taken to be the weight of the object.

The modern laboratory is of course equipped with balances which reflect the technological advances that have taken place over the years. Nearly all modern laboratory balances are of the single pan variety. The basic principle of the single-pan analytical balance is shown in Figure 2.2. In this design, there is only one pan. A permanent, constant counterbalancing weight is in place across a fulcrum from the pan. The object to be weighed is placed on the pan and, along with a series of removable weights on the same side, are

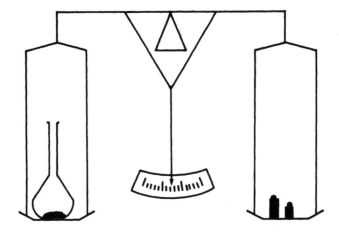

FIGURE 2.1   The basic concept of the device known as a "balance." A fulcrum, or knife-edge, supports a beam across which two "pans" are balanced; one containing the object being weighed and the other containing known weights.

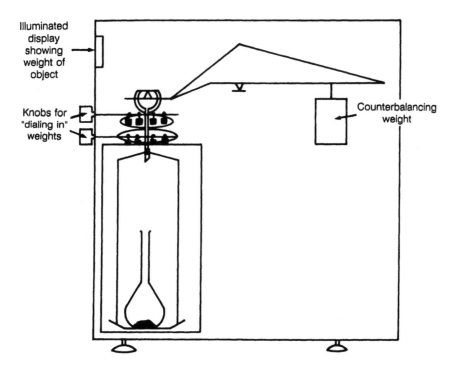

FIGURE 2.2   A schematic diagram of a single-pan balance.

made to balance the constant weight on the other side. When a heavy object is placed on the pan, some of the removable weights are lifted via an external control so as to achieve a balance with the counterbalancing weight. We will discuss these in more detail in the next subsection (Section 2.2.3).

Different examples of laboratory work require different degrees of accuracy. Thus, there are a variety of balance designs available which reflect this need for varying accuracy. Some balances are accurate to the nearest gram, some are accurate to the nearest tenth or hundredth of a gram, some are accurate to the nearest milligram, and still others are accurate to the nearest tenth and hundredth of a milligram. We now proceed to describe some of the more popular designs, and our discussions will begin with a shorter treatment of some of the less accurate balances and end with a detailed treatment of the important electronic analytical balances, which are very accurate measuring devices.

2.2.2 The Less Accurate Balances

Figure 2.3 depicts two common balances that can be described as "less accurate" or "auxiliary." These are used when the weight being measured does not necessarily need to be accurate. (See Chapter 1 for a discussion of when to be accurate and when not to be accurate.) The multiple-beam balance (Figure 2.3a) is accurate to the nearest 0.01 g and resembles a balance commonly found in a doctor's office to measure a person's weight. The basic design here is more like the double pan variety described briefly above in the sense that weights are added on a counterbalancing arm to balance the object on the pan, rather than removing weights from the same side as the pan.

Electronic top-loading balances (Figure 2.3b) have variable accuracies depending on the model purchased. The one illustrated is an Ohaus Precision Standard balance, which is accurate to the nearest 0.01 g. Other such balances are accurate to the nearest 0.1 and 0.001g. The electronic top loaders often have a "tare" feature. A chemical can conveniently be weighed on a piece of weighing paper, for example, without having to determine the weight of the paper. "Taring" means that the balance is simply zeroed with the weighing paper on the pan. These balances are not based on balance/counterbalance, but rather utilize a torsion mechanism.

a

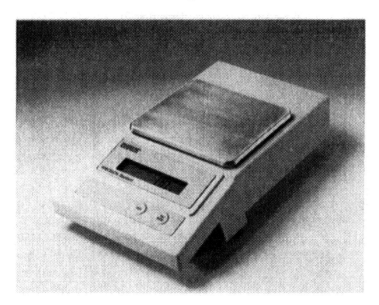

b

FIGURE 2.3   (a) A multiple-beam balance. (b) An electronic top-loading balance. (Courtesy of Ohaus Corporation, Florham Park, NJ.)

2.2.3 The Electronic Analytical Balances

The balances that are used to achieve the highest degree of accuracy in the analytical laboratory are called "analytical" balances and are accurate to either 0.1 or 0.01 mg. The modern laboratory utilizes single-pan analytical balances almost exclusively for such accuracy. Such balances can be either of the balance/counterbalance variety or the torsion variety.

Figure 2.4 shows a typical modern analytical balance. Notice that it is a single-pan balance with the pan enclosed. The chamber housing the pan has transparent walls for easy viewing. Sliding doors on the right and left sides make the pan accessible for loading and handling samples. The actual weight measurement is displayed via a digital readout on the face of the instrument. The counterbalancing mechanism, if it has one, is hidden in the upper portion of the instrument and is usually accessible by removing the cover.

The design of the readout and the exact weighing procedure varies with the age of the balance and the manufacturer. The specific techniques, such as the use of a "pan arrest" lever, for obtaining the readout are easy to learn, however, through demonstration and practice and, therefore, will not be discussed in detail here.

The step-by-step instructions which follow are written to be as generic as possible. The most important point is that the operator keep in mind that the analytical balance is extremely sensitive and extremely accurate and, therefore, must be handled carefully and correctly. Also, this discussion presumes that, if necessary, the sample has been dried (such as in an oven) and has been kept dry (by storing in a desiccator) prior to weighing.

Step 1.   Some preliminary considerations are important. First, the sample to be weighed must be at room temperature. If it is not at room temperature, it must first be allowed to cool. The reason for this is that a warm or hot object placed on the pan in the enclosed chamber can create air currents that can buoy up the pan and cause an erroneous weight reading. Second, whenever a weight is being read on the readout, the sliding doors must be closed. Again, this is to avoid the effect of air currents on the weight measurement. Third, the pan and the floor of the chamber must be swept free of chemical debris from prior spills. A camel-hair brush should be kept near the

FIGURE 2.4   A typical modern analytical balance. (Courtesy Ohaus Corporation, Florham Park, NJ.)

balance for this purpose. Fourth, care should be taken to avoid bumping the balance or its support while the measurement is taken. This would almost always cause the reading to change or vibrate and could also be damaging to the internal instrument parts. Fifth, the balance should be level. With most balances, a leveling method is built in complete with a leveling bubble and vertically positionable legs.

Step 2.   If the absolute weight of an object on the pan is needed, such as a crucible weight, the balance must be zeroed. This is a calibration step in which, when there is nothing on the pan, the readout is made to read zero. All designs of analytical balances have a "zeroing control" for this purpose. With nothing on the pan, the zeroing control is adjusted such that the readout reads zero even to the fourth or fifth decimal place. If the absolute weight of an object on the pan is not needed, such as when weighing by difference or in taring, the zero step is not absolutely necessary.

Step 3.   The object placed on the pan must be kept free of fingerprints, or other interfering substances, that could add weight and give an erroneous result. The use of gloves or finger cots is recommended. When weighing an object to the nearest tenth of a milligram, seemingly insignificant fingerprints can cause quite a significant error. Tongs can also be helpful here.

Step 4.   The object is placed on the pan and the weight obtained. As indicated earlier, balances vary in the mechanism for doing this. Some have a "preweigh" mechanism, and most require pan arrest while dialing in weights or when adding and removing objects from the pan. Most require the operator to "dial in" the weights so as to achieve a null of some type. Modern torsion type balances have a digital readout that automatically displays the weight without "dialing it in."

Step 5.   When finished, dial all weight controls to zero if necessary, clean up any spills, arrest the pan if required, and turn off any switches.

2.3 GLASSWARE

The accurate measurement of volumes of solutions is a very important measurement in analytical chemistry. It is important for volumes of titrants in titrimetric analysis to be measured accurately. Final solution volumes must be measured accurately when preparing solutions of accurate concentration. A variety of reasons exist for accurately transferring a volume of a solution from one vessel to another. It should not be surprising that an analytical chemist needs to be well versed in the selection and proper use of volumetric glassware.

There are basically three types of volume measuring devices in common use. These are the volumetric flask, the pipet, and the buret. Let us study the characteristics of each type individually.

2.3.1 The Volumetric Flask

Let us first deal with the container that is typically used for accurate solution preparation. We emphasize the word "accurate" here to distinguish it from solution preparation procedures which do not need to be accurate. In these latter cases, ordinary beakers, Erlenmeyer flasks, graduated cylinders, and uncalibrated bottles can be used. (See Chapter 1, Section 1.5, for a discussion of when accuracy is important and when it is not.) The container of choice for accurate solution preparation is the volumetric flask. A drawing of the volumetric flask is given in Figure 2.5a.

This flask is characterized by a large base that tapers into a very narrow neck on which a single calibration line is affixed. This calibration line is affixed so that the indicated volume is *contained* rather than delivered. Accordingly, the legend "TC" is imprinted on the base of the flask, thus marking the flask as a vessel "to contain" the volume indicated as opposed to "to deliver" a volume. The reason this imprint is important is that a contained volume is different from a delivered volume, since a small volume of solution remains adhering to the inside wall of the vessel and is not delivered when the vessel is drained. If a piece of glassware is intended to deliver a specified volume, the calibration must obviously take this small volume into account in the sense that it *will not* be part of the *delivered* volume. On the other hand, if a piece of glassware is not intended to deliver a specified volume, but rather to contain the volume, the calibration must take this small volume into account in the sense that it *will* be part of the *contained* volume. Other pieces of glassware, namely, most pipets and all burets, are "TD" vessels, meaning they are calibrated "to deliver." Figure 2.5b shows a closeup of the base of a volumetric flask clearly showing the "TC" imprint (just below the flask's capacity — 500 mL).

Notice the other markings on the base of the flask in Figure 2.5b. The imprint "A" refers to flasks (and pipets and burets as well) that have undergone more stringent calibration procedures at the factory (Class A flasks). Such flasks are more accurate (and also more expensive) than flasks that do not have this marking. The imprint "20°C" indicates that the flask is calibrated to contain the indicated volume when the temperature is 20°C, which is a standard temperature of calibration. This marking is needed since the volume of liquids and liquid solutions changes slightly with temperature. For highest accuracy, the temperature of the contained fluid should be adjusted to 20°C.

One final marking on the base of the flask is the "19" designation. This refers to the size of the tapered top on the flask. The stopper that is used can be either a ground glass stopper, to match the flask

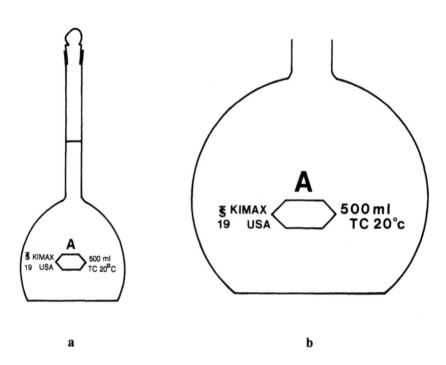

FIGURE 2.5   (a) A drawing of a volumetric flask. (b) A closeup of the base of a volumetric flask clearly showing the information imprinted there.

opening, or it can simply be a plastic tapered stopper, which can also be used in a ground glass opening. It has an identical numerical imprint on it ("19" in this case) and such number designations on these two items must match to indicate that the stopper is the correct size. The stopper is not necessarily a tapered stopper as in this example. It is fairly common for the stopper to be a "snap cap." In this case, the top of the flask is not tapered nor is it a ground glass opening. Rather, it has an unusually large lip around the opening, over which the snap cap is designed to seal. A number designation, however, is used as with the tapered opening to indicate the size of the opening and the size of the cap required. This number is found on both the flask base and the cap as in the other design.

Volumetric flasks are manufactured in a variety of sizes. Since there is only a single calibration line, the use is limited to rather common volumes: 5, 10, 25, 50, 100, 200, 250, and 500 mL, and 1 and 2 L. One problem with the volumetric flask exists because of its unique shape, and that is the difficulty in making prepared solutions homogeneous.

When the flask is inverted and shaken, the solution in the neck of the flask is not agitated. Only when the flask is set upright again is the solution drained from the neck and mixed. A good practice is to invert and shake at least a dozen times to ensure homogeneity.

Volumetric flasks should *not* be used to prepare solutions of reagents that can etch glass (such as sodium hydroxide and hydrofluoric acid), since if the glass is etched, its accurate calibration is lost. Volumetric flasks should *not* be used for storing solutions. Their purpose is to prepare solutions accurately. If they are used for storage, then they are not available for their intended purpose. Finally, volumetric flasks should *not* be used to contain solutions when heating or performing other tasks for which their accurate calibration serves no useful purpose. There are plenty of other glass vessels to perform these functions.

2.3.2 The Pipet

As indicated above, most pipets are pieces of glassware that are designed to deliver (TD) the indicated volume. Pipets come in a variety of sizes and shapes. The most common is probably the "volumetric" or "transfer" pipet shown in Figure 2.6. This pipet, like the volumetric flask, has a single calibration line. It can thus be used in delivering only rather common volumes — 5, 10, 15 mL, etc. The correct use of a volumetric pipet is outlined in Table 2.1. The sequence of events is described with less detail as follows. The bottom tip of the pipet is placed into the solution to be transferred. A pipet bulb is evacuated, placed over the top of the pipet, and slowly released. The solution is drawn up into the pipet as a result of the vacuum. When the level of the solution has risen above the calibration line, the bulb is quickly removed from the pipet and replaced with the index finger. Next, the tip of the pipet is removed from the solution and wiped with a towel. The pressure exerted with the index finger is then used to adjust the bottom of the meniscus to coincide with the calibration line. The tip of the pipet is placed into the receiving vessel, and the finger is released. The solution is thus drained into the receiving vessel. When the solution has completely drained from the pipet, the bottom tip should be placed against the wall of the receiving vessel so that the last bit of solution will drain out. A small drop of

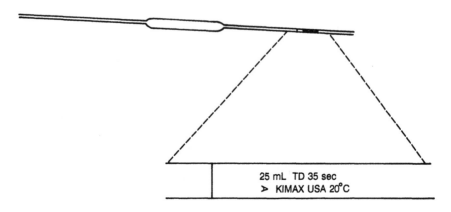

25 mL TD 35 sec
➤ KIMAX USA 20°C

FIGURE 2.6   A volumetric pipet and the top of a class "A" volumetric pipet showing the markings described in the text.

the liquid will remain in the pipet at this point. The pipet should be given a half-turn (twist) and then removed. *Under no circumstances should this last drop in a volumetric pipet be blown out with the bulb.* The volumetric pipet is not calibrated for "blow-out."

For precise work, volumetric pipets that are labeled as Class A have a certain time in seconds imprinted near the top (see Figure 2.6) which is the time that should be allowed to elapse from the time the finger is released until the pipet is given the half-turn and removed. The reason for this is that the film of solution adhering to the inner walls will continue to slowly run down with time, and the length of time one waits to terminate the delivery thus becomes important. The intent with Class A pipets, then, is to take this "run-down" time into account by terminating the delivery in the specified time. After this specified time has elapsed, the pipet is touched to the wall of the receiving flask, given the half-turn, and removed.

Several additional styles of pipets other than the volumetric pipet are in common use. These are shown in Figure 2.7. Pipets that have graduation lines, much like a buret, are called "measuring" pipets. They are used whenever odd volumes are needed. There are two types of measuring pipets: the Mohr pipet and the serological pipet. The difference is whether or not the calibration lines stop short of the tip (Mohr Pipet) or go all the way to the tip (serological pipet). The serological pipet is better in the sense that the meniscus need be read only once since the solution can be allowed to drain completely out. In this case, the last drop of solution is blown out with the pipet bulb. With the Mohr pipet, however, the meniscus must be read twice: once before the delivery and again after the delivery is complete. The

Table 2.1   The Sequence of Events Involved in Transferring a Volume of Solution with a Volumetric Pipet (See Text for More Discussion)

Step 1.	If the outside of the pipet is wet, dry it with a paper towel first, especially the tip.
Step 2.	Evacuate the pipet bulb (by squeezing).
Step 3.	Seat the bulb opening over the top opening of the pipet.
Step 4.	If the pipet is wet on the inside with a foreign liquid, immerse the tip of the pipet into the solution to be delivered while simultaneously releasing the bulb slightly to immediately draw the solution in to about half the pipet's capacity. Empty by inverting and draining into a sink through the top. Repeat this rinsing step at least three times. If the inside of the pipet is dry, proceed with Step 5.
Step 5.	Fill the pipet to well past the calibration line by releasing the squeezing pressure, as in Step 4. Reevacuate the bulb if necessary.
Step 6.	Quickly remove the bulb and seal the top of the pipet with the index finger.
Step 7.	Keeping the index finger in place, remove the tip from the solution and wipe with a towel. To avoid contamination from the towel, tilt the pipet to a 45° angle so that a small volume of air is drawn into the tip before wiping.
Step 8.	Slowly release the finger to adjust the meniscus to the calibration line.
Step 9.	Touch the tip to the *outside* of the receiving vessel so as to remove a drop of the solution that may be suspended there.
Step 10.	Place the tip into the mouth of the receiving vessel and completely release the finger.
Step 11.	When the draining is complete, touch the tip to the inside wall of the vessel and give it a half-twist.

solution flow out of the pipet must be halted at the correct calibration line, and the error associated with reading a meniscus is thus doubled. The delivery of 4.62 mL, for example, is done as in Figure 2.8a with a Mohr pipet, but as in Figure 2.8b, it is with a serological pipet.

It should be stressed that with the serological pipet, every last trace of solution capable of being blown out must end up inside the receiving vessel. Some analysts find that this is more difficult and perhaps introduces more error than reading the meniscus twice with a Mohr pipet. For this reason, these analysts prefer to use a serological pipet as if it were a Mohr pipet. It is really a matter of personal preference. A double or single frosted ring circumscribing the top of the pipet (above the top graduation line) indicates the pipet is calibrated for blow-out. "Disposable" pipets are most often of the serological blow-out type. They are termed disposable because the calibration lines are not necessarily permanently affixed to the outside wall of the pipet. The calibration process is thus less expensive, resulting in a less expensive product which can be discarded after use.

One of two pipets that is calibrated "TC" is called the lambda pipet

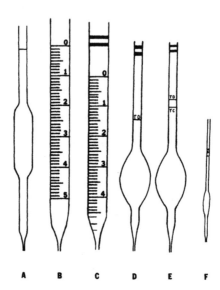

FIGURE 2.7   Some types of pipets: (A) volumetric, (B) Mohr, (C) serological, (D) Ostwald-Folin, (E) duopette, and (F) lambda.

(Figure 2.7). Such a pipet is used to transfer unusually viscous solutions such as syrups, blood, etc. With such solutions, the thin film remaining inside would represent a significant nontransferred volume which translates into a significant error by normal "TD" standards. With the lambda pipet, the calibration line is affixed at the factory so that every trace of solution contained within is transferred by flushing the solution out with a suitable solvent. Thus, the pipetted volume is contained within and then quantitatively flushed out. Such a procedure would actually be acceptable with any "TC" glassware, including the volumetric flask. Obviously, diluting the solution in the transfer process must not adversely affect the experiment. Some pipets have both a TC line and a TD line and are called "duopettes" (Figure 2.7). Ostwald-Folin pipets (Figure 2.7) have a single line but are calibrated for blow-out.

2.3.3 Pipetting Devices

In addition to the several designs of pipets just described, various pipetting "devices" for measuring solution volumes from 0 to 1 mL

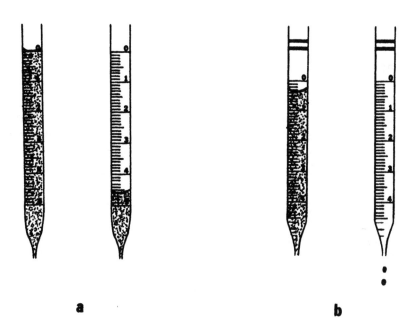

FIGURE 2.8   The delivery of 4.62 mL of solution (a) with a Mohr pipet (the meniscus is read twice) and (b) with a serological pipet (the meniscus is read once [at 0.38] and the solution blown out).

have been invented for laboratory use and have become very popular. These "pipettors" employ a bulb concealed within a plastic fabricated body, a spring loaded push-button at the top, and a nozzle at the bottom for accepting a plastic disposable tip. They may be fabricated for either single or variable volumes. In the latter case, a ratchet-like device with a digital volume scale is present below the push-button. The desired volume may be "dialed in" prior to use. An example is shown in Figure 2.9.

2.3.4 The Buret

The buret has some unique attributes and uses. One could call it a specialized graduated cylinder, having graduation lines that increase from top to bottom, with a usual accuracy of ± 0.01 mL, and a stopcock at the bottom for dispensing a solution. There are some

FIGURE 2.9   An Eppendorf pipettor with variable volume capacity.

variations in the type of stopcock that warrant some discussion. The stopcock itself, as well as the barrel into which it fits, can be made of either glass or Teflon™.* Some burets have an all glass arrangement, some have a Teflon stopcock and a glass barrel, and some have the entire system made of Teflon. The three types are pictured in Figure 2.10.

When the all-glass system is used, the stopcock needs to be lubricated so it will turn with ease in the barrel. There are a number of greases on the market for this purpose. Of course, the grease must be inert to chemical attack by the solution to be dispensed. Also, the amount of grease used should be carefully limited so that excess grease does not pass through the stopcock and plug the tip of the buret. (Any material stuck in a buret tip can usually be dislodged with a fine wire inserted from the bottom when the stopcock is open and the buret full of solution.) The Teflon stopcocks are free of this lubrication problem. The only disadvantage is that the Teflon can become deformed, and this can cause leakage.

The correct way to position one's hands to turn the buret's stopcock during a titration is shown in Figure 2.11. The natural tendency with this positioning is to pull the stopcock in as it is turned. This will prevent the stopcock from being pulled out, causing the titrant to bypass the stopcock. The other hand is free to swirl the flask as shown.

2.3.5 Cleaning and Storing Procedures

The use of clean glassware is of utmost importance when doing a chemical analysis. In addition to the obvious need of keeping the solution free of contaminants, the walls of the vessels, particularly the transfer vessels (burets and pipets), must be cleaned so the solution will flow freely and not "bead up" on the wall as the transfer is

---

*   Registered Trademark of E.I. DuPont Nemours and Company, Inc., Wilmington, DE.

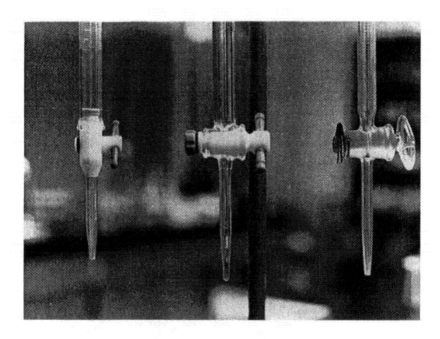

FIGURE 2.10   The three types of stopcocks: (left) Teflon stopcock, Teflon barrel, (middle) Teflon stopcock, glass barrel, and (right) glass stopcock, glass barrel. (From Kenkel, J., *Analytical Chemistry for Technicians,* Lewis Publishers, Inc., Chelsea, MI, 1988. With permission.)

performed. If the solution beads up, it is obvious that the pipet or buret is not delivering the volume of solution that one intends to deliver. It also means that there is a film of grease on the wall which could introduce contaminants. The analyst should examine, clean, and reexamine his/her glassware in advance so that the free flow of solution down the inside of the glassware is observed. For the volumetric flask, at least the neck must be cleaned in this manner so as to ensure a well-formed meniscus.

Of course, the next question is "What cleaning procedures are used?" For most cleaning requirements, ordinary soap and water used with a brush where possible is sufficient. A commercial laboratory soap called Alconox™ is one product often used. Other excellent phosphate-free detergents are available.* With burets, a cylindri-

---

* Registered trademark of Alconox, Inc., New York, NY.

FIGURE 2.11   The correct way to position one's hands for a titration. (From Kenkel, J., *Analytical Chemistry for Technicians*, Lewis Publishers, Inc., Chelsea, MI, 1988. With permission.)

cal brush with a long handle (buret brush) is used to scrub the inner wall. With flasks, a bottle or test tube brush is used to clean the neck. Also, there are special bent brushes available to contact and scrub the inside of the base of the flask.

Pipets pose a special problem. Brushes cannot be used because of the shape of some pipets and the narrowness of the openings. In this case, if soap is to be used, one must resort to soaking with a warm soapy water solution for a period of time proportional to the severity of the particular cleaning problem. Commercial soaking and washing units are available for this latter technique. Soap tablets are manufactured for such units and are easy to use.

For pipets and for difficult cleaning problems for other pieces of glassware, special cleaning solutions, which chemically break down

Table 2.2   Recipes for Preparing Glassware Cleaning Solutions

Chromic acid, Method 1	Dissolve 90–100 g of sodium or potassium dichromate in 450 mL of water and add 80 mL of concentrated sulfuric acid. The solution will become a semisolid red mass. Add just enough sulfuric acid to dissolve this mass.
Chromic acid, Method 2	Add the contents of a 25-mL bottle of Monostat "Chromerge" (available from chemical products suppliers) to a standard 9-lb bottle of concentrated sulfuric acid. Add approximately 5 mL at a time and shake well after each addition.
Alcoholic KOH	Dissolve 105 g of KOH in 120 mL of water. Add 1 L of 95% ethanol (or propanol) and shake well. This solution will etch glass to a small extent and therefore should not be left in contact with volumetric glassware or ground glass joints for more than about 15 min at a time.

Reprinted from Kenkel, J., *Analytical Chemistry for Technicians*, Lewis Publishers, Inc., Chelsea, MI, 1988. With permission.

greasy films through soaking for a period of time, are used. One commonly-used such cleaning solution is chromic acid. This is a solution of concentrated sulfuric acid and potassium dichromate. Another is a solution of potassium hydroxide in ethanol or propanol. Recipes for preparing these solutions are given in Table 2.2. These are very tough on serious cleaning problems.

Safety should be stressed in the use of these cleaning solutions. The highly corrosive nature of the ingredients makes it imperative to prevent spills and splashes on one's person. In addition to normal safety gear (safety glasses, eye wash, shower, first aid kit), one should also have solutions of weak acids (e.g., acetic acid, citric) and weak bases (e.g., sodium carbonate) handy to neutralize spills quickly. It is highly recommended that use of these cleaning solutions be restricted to a good fume hood. Spent cleaning solutions should *not* be poured down the drain, but disposed of like any other hazardous waste.

Once the glassware has been cleaned (by whatever method), one should also take steps to keep it clean. One technique is to rinse the items thoroughly with distilled water and then dry them in an oven. Following this, they are cooled and stored in a drawer. For shorter time periods, it may be convenient to store them in a soaker under distilled water. This prevents their possible recontamination during dry storage. When attempting to use a soaker-stored pipet or buret, however, it must be remembered that the thin film of water is present on inner walls, and this must be removed by rinsing with the solution to be transferred, being careful not to contaminate the solution in the process.

2.4 REAGENTS AND METHODS FOR SAMPLE PREPARATION

For the vast majority of chemical analysis schemes, the samples to be analyzed must be dissolved in a liquid solution. Many samples, when they first arrive in the laboratory, are either undissolved solids or liquids from which the analyte must be removed and dissolved in another solvent before proceeding. The first operation often performed on a sample is therefore either total dissolution or an extraction. There is a broad array of substances available, and a few special techniques used for these operations and their use depends on the nature of the sample and whether complete dissolution or an extraction is needed. We now present a discussion of some of the more common laboratory reagents and extraction techniques and examples of their use.

2.4.1 Water, Concentrated Acids, and Bases

Water — It should come as no surprise that ordinary water can be an excellent solvent for many samples. Due to its extremely polar nature, water will dissolve most substances of likewise polar or ionic nature. Obviously, then, when samples are composed solely of ionic salts, water would be an excellent choice. An example might be the analysis of a commercial table salt for sodium iodide content. A list of "solubility rules" for ionic compounds in water can be found in Table 2.3.

Hydrochloric Acid — Strong acids are used frequently for the purpose of sample dissolution when water will not do the job. One of these is hydrochloric acid, HCl. Concentrated HCl is actually a saturated solution of hydrogen chloride gas, and the fumes are very pungent. Such a solution is 38.0% HCl and about 12 M. Hydrochloric

Table 2.3 Solubility Rules for Some Common Inorganic Compounds in Water

Compound Class	Soluble?	Exceptions
Nitrates	Yes	None
Acetates	Yes	Silver acetate sparingly soluble
Chlorides	Yes	Chlorides of Ag, Pb, and Hg are insoluble
Sulfates	Yes	Sulfates of Ba, Pb are insoluble; sulfates of Ag, Hg, and Ca are slightly soluble
Carbonates	No	Carbonates of Na, K, and NH_4 are soluble
Phosphates	No	Phosphates of Na, K, and NH_4 are soluble
Chromates	No	Chromates of Na, K, NH_4, and Mg are soluble
Hydroxides	No	Hydroxides of Na, K, NH_4 are soluble; hydroxides of Ba, Ca, and Sr are slightly soluble
Sulfides	No	Sulfides of Na, K, NH_4, Ca, Mg, and Ba are soluble
Sodium salts	Yes	Some rare exceptions
Potassium salts	Yes	Some rare exceptions
Ammonium salts	Yes	Some rare exceptions
Silver salts	No	Silver nitrate, perchlorate; silver acetate and sulfate are sparingly soluble

acid solutions are used especially for dissolving metals, metal oxides, and carbonates not ordinarily dissolved by water. Examples are iron and zinc metals, iron oxide ore, and the metal carbonates of which the scales in boilers and humidifiers are composed. Being a strong acid, it is very toxic and must be handled with care.

Sulfuric Acid — An acid that is considered a stronger acid than HCl in many respects is sulfuric acid, H_2SO_4. When sulfuric acid contacts clothing, paper, etc., one can see an almost instantaneous reaction — paper towels turn black and disintegrate, and clothing fibers become weak and holes readily form. Concentrated sulfuric acid is about 96% H_2SO_4 (the remainder being water) or about 18 M and is a clear, colorless, syrupy, dense liquid. It reacts violently with water, evolving much heat, and so water solutions of sulfuric acid must be prepared cautiously, often to include a means of cooling the container. Its sample dissolution application is limited mostly to organic material, such as vegetable plants. It is the solvent of choice for the Kjeldahl analysis (Chapter 4) for such materials as grains and products of grain processing. It is also used to dissolve aluminum and titanium oxides on airplane parts.

Nitric Acid — Another acid that has significant application is nitric acid, HNO_3. This acid is also very dangerous and corrosive and is aptly referred to as an oxidizing acid. This means that a reaction other than hydrogen gas displacement (as with HCl) occurs when it contacts metals. Frequently, oxides of nitrogen form in such a reaction, and noxious brown, white, and colorless gases are evolved. Concentrated HNO_3 is 70% HNO_3 (16 M) and is used for applications where a strong acid with additional oxidizing power is needed. These include metals such as silver and copper, as well as organic materials such as in a wastewater sample. Nitric acid will turn skin yellow after only a few seconds of contact.

Hydrofluoric Acid — An acid that has some very useful and specific applications, but is also very dangerous, is hydrofluoric acid, HF. This acid reacts with skin in a way that is not noticeable at first, but becomes quite serious if left in contact for a period of time. It has been known to be especially serious if trapped against the skin and after diffusing under fingernails. Treatment of this is difficult and painful. Concentrated HF is about 50% HF (26 M). It is an excellent solvent for silica (SiO_2)-based materials such as sand, rocks, and glass. It can also be used for stainless steel alloys. Since it dissolves glass, it must be stored in plastic containers. This is also true for low pH solutions of fluoride salts.

Perchloric Acid — Another important acid for sample preparation and dissolution is perchloric acid, $HClO_4$. It is an oxidizing acid like HNO_3, but is considered to be even more powerful in that regard when hot and concentrated. It is most useful for more difficult organic samples, such as leathers and rubbers. It can also be used for stainless steels and other more stable alloys. Commercial $HClO_4$ is 72% $HClO_4$ (12 M). It is a very dangerous acid and should only be used in a fume hood designed for the collection of its vapors. Contact with alcohols, including polymeric alcohols, such as cellulose, and other oxidizable materials, should be avoided due to the potential for explosions.

"**Aqua Regia**" — An acid mixture that is prepared by mixing one part concentrated HNO_3 with three parts concentrated HCl is called "aqua regia." This mixture is among the most powerful dissolving agents known. It will dissolve the very noble metals (gold and platinum), as well as the most stable of alloys.

Sodium Hydroxide and Ammonium Hydroxide — Bases such as sodium hydroxide and ammonium hydroxide, while quite commonly used in the analytical laboratory, are used sparingly in sample preparation. The high pH of such solutions often does not help with sample dissolution and in fact may hinder it. A case in point is metals. While many metals are quite soluble at low pH values, many form highly insoluble hydroxides at basic pH values. Thus, bases cannot be used to dissolve samples containing metals (unless it is an extraction in which the metal is not the analyte — see following subsection). However, a hydroxide precipitate may be a *desired* product, such as in a scheme for the separation of a metal from an analyte by its precipitation. When high pH values are desirable, both sodium hydroxide, a strong base, and ammonium hydroxide, a weak base, have a usefulness.

2.4.2 Sample Preparation By Extraction

In cases in which it is not practical or necessary to dissolve an entire sample or in which it is easy to affect a "separation" of the analyte by selective dissolution, an extraction procedure is used. This technique is described in some detail in Chapter 8.

Briefly, the sample can be either a solution (typically a water solution) of the analyte or a solid material, e.g., soil. When the sample is a water solution, solvent extraction refers to an experiment in which an organic solvent, immiscible with the sample, is brought into intimate contact with the sample (in a "separatory funnel" — see Chapter 8) in such a way that the substance of interest is at least partially removed (extracted) from the water and dissolved in the solvent.

When the sample is a solid, the extracting solvent is often an aqueous solution of an inorganic compound, such an acid, although it can also be an organic solvent. Here again, the sample is brought into intimate contact with the extracting solvent by shaking in the same container (usually after grinding to a fine powder) or by repeated contact with fresh solvent in an apparatus called a Soxhlet extractor (Chapter 8). With both of these methods, the analyte is at least partially separated from the sample and dissolved in this solvent, while most other materials remain undissolved. This method is acceptable since the undissolved materials play no role at all in the analysis and, in fact, may contain interfering substances which are best left undissolved. The analysis can then proceed on the sample "extract" as with any other sample solution. Please refer to Chapter 8 for more details of the extraction technique.

Following is a discussion of the properties of common organic solvents used in the laboratory for extractions and other applications.

n-Hexane — n-Hexane is nonpolar and an alkane, C_6H_{14}. Its solubility in water is virtually nil. It is less dense than water (density = 0.66 g/mL); thus it would be the top layer in a separatory funnel with a water solution. It is obviously a poor solvent for polar compounds, but is very good for extracting traces of nonpolar solutes in water samples. It is used as a solvent for UV spectrophotometry. It is highly flammable and has a low toxicity level. It can be used as the mobile phase in normal phase chromatography.

Acetonitrile — Acetonitrile is a polar solvent, $CH_3C{\equiv}N$. It is completely miscible with water. It is used for the extraction of polar components of solid samples, such as the extractions of chlorinated pesticide residues from vegetables, and also as a mobile phase for reverse phase chromatography. It can be used as a solvent for UV and NMR work. It is flammable and toxic and has an unpleasant odor. Breathing acetonitrile vapors should be avoided.

Methylene Chloride — This solvent is a somewhat polar solvent also known as dichloromethane, CH_2Cl_2. Its solubility in water is about 2%. It is denser than water (density = 1.33 g/mL); thus it would be the bottom layer when used with a water solution in a separatory funnel. It may form an emulsion when shaken in a separatory funnel with water solutions. It is used for extractions, as well as for infrared (IR) and NMR solvents. It is not flammable and is considered to have a low toxicity level.

Acetone — Acetone (CH_3COCH_3) is a volatile, highly flammable liquid that has intermediate polarity and is completely miscible with water. It is used as a solvent for both polar and nonpolar substances, and is therefore used often as a cleaning solvent for glassware and as a degreaser. Its high volatility makes it useful for rapid drying of

glassware. It can be used as a solvent for NMR work. It is occasionally used to extract components from solid samples. It is considered nontoxic, but prolonged breathing of vapors should be avoided.

Benzene — Benzene (C_6H_6) is a nonpolar aromatic liquid insoluble in water. It can be used as a solvent in NMR work. For extraction work, it has been used for extraction of nonpolar components of water samples, but is prone to emulsion formation. Its density is 0.87 g/mL. It is highly flammable and toxic. It is especially noted as a cancer-causing carcinogen. Breathing of vapors and skin contact should be avoided. Due to its toxicity, it has been replaced in most laboratory applications with toluene.

Toluene — Toluene (C_6H_5–CH_3) is also a nonpolar aromatic liquid insoluble in water. It is flammable, but less toxic than benzene. Its density is 0.87 g/mL and is slightly soluble in water (0.47 g/L). It is sometimes used as a mobile phase in normal phase liquid chromatography and as an extraction solvent for nonpolar sample components.

Methanol — Methanol (CH_3–OH) is a polar organic solvent completely miscible with water and most other solvents. It is a flammable and poisonous liquid. It is useful as a solvent for UV and NMR work, as well as a mobile phase for reverse phase liquid chromatography. Like acetone, it is sometimes used to quick-dry laboratory glassware, since it is volatile. It is poisonous when ingested, but inhalation should also be avoided.

Diethyl Ether — Diethyl ether (CH_3CH_2–O–CH_2CH_3), also frequently referred to as "ethyl ether" and as "ether," is a mostly nonpolar organic liquid that is highly volatile and extremely flammable, a dangerous combination. Also, explosive peroxides form with time. Precautions regarding storage to discourage peroxide formation include storage in metal containers in explosion-proof refrigerators. It should be disposed of after about nine months of storage.

Ether is only slightly soluble (75 g/L) in water. Given its slight polarity, it is useful to extract nonpolar components from water. Its density is 0.71 g/mL. Since it is volatile, it is easily evaporated from extraction fractions.

Chloroform — Chloroform (trichloromethane, $CHCl_3$) is a nonflammable organic liquid with low miscibility with water (10 g/L). It is useful as an extracting liquid for water samples; as a solvent for UV, IR, and NMR work; and as a mobile phase for HPLC. It is very toxic and should be avoided when another solvent would do as well. Vapors should not be inhaled and contact with the skin should be avoided.

2.4.3 Fusion

For extremely difficult samples, a method called "fusion" may be employed. Fusion is the dissolving of a sample using a molten inorganic salt generally called a "flux." This flux dissolves the sample and, upon cooling, results in a solid mass that is then soluble in a liquid reagent. The dissolving action is mostly due to the extremely high temperatures (usually 300 to 1000°C) required to render most inorganic salts molten.

Additional problems arise within fusion methods, however. One is the fact that the flux must be present in a fairly large quantity in order to be successful. The measurement of the analyte must not be affected by this large quantity. Also, while a flux may be an excellent solvent for difficult samples, it will also dissolve the container to some extent, creating contamination problems. Platinum crucibles are commonly used, but nickel, gold, and porcelain have been successfully used for some applications.

Probably the most common flux material is sodium carbonate. It may be used by itself (melting point 851°C) or in combination with other compounds, such as oxidizing agents (nitrates, chlorates, and peroxides). These may be used to dissolve silicates and silica-based samples, as well as samples containing alumina (Al_2O_3).

For dissolving particularly difficult metal oxides, the acidic flux potassium pyrosulfate ($K_2S_2O_7$) may be used. The required temperature for this flux is about 400°C.

2.4.4 Reagent Quality

It is of utmost and obvious importance in an analytical laboratory to be able to trust all reagents used for both the sample preparation schemes, which have been outlined in this section, as well as for other analytical preparations. A stringent quality assurance program for such reagents is necessary. This means that the quality of all inventoried, and newly purchased chemicals as well, must be assured. It also means that the integrity of the various solutions prepared and samples gathered for analytical purposes in the laboratory must be maintained.

First, chemicals purchased and/or stored will have a designation of their purity on the label. This designation will likely be one of the following.*

Primary Standard	This grade designates specially manufactured analytical reagent of exceptional purity for standardizing solutions and preparing reference standards.
Reagent (ACS)	Maximum limits of purity for most commonly used reagents (mostly inorganic) have been established by the Committee on Analytical Reagents of the American Chemical Society. Whenever a specification exists and a product is produced in conformity with it, the bottle is so labeled.
Reagent	When the American Chemical Society has not developed specifications for a specific reagent, the manufacturer establishes its own standards, and the maximum limits of allowable impurities are shown on the labels of these reagents.
C.P.	"Chemically Pure" grades of chemicals are offered by manufacturers. They meet or exceed U.S.P. or N.F. requirements, but are lower grade than "Reagent" or "Reagent, ACS" chemicals.
U.S.P.	Chemicals labeled U.S.P. meet the requirements of the U.S. Pharmacopeia. Generally of interest to the pharmaceutical profession, these specifications may not be adequate for reagent use. This designation and the N.F. designation are now the same.
N.F.	Chemicals labeled N.F. meet the requirements of the National Formulary. Chemicals ordered with an N.F. label may not be useful for re-

---

*   These designations have been reprinted, in part, from the *Modern Chemical Technology Guidebook*, Revised Edition, 1972, 157. With permission. Copyright by the American Chemical Society, Washington, D.C.

	agents. It will be necessary to check the National Formulary in each case. This designation and the U.S.P. designation are now the same.
Practical	This grade designates chemicals of sufficiently high quality to be suitable for use in some syntheses. Organic chemicals of practical grade may contain small amounts of intermediates, isomers, or homologs.
Purified	This is a grade of chemical for which care has been exercised by the manufacturer to offer a product that is physically clean and of good quality, but not meeting Reagent ACS, Reagent, U.S.P., or C.P. standards.
Technical	This is a grade of chemical generally suitable for industrial use. Purity is not specified and is generally determined by on-site analysis.
Spectro Grade	This is a designation for organic solvents which have been prepared for use in UV or IR spectroscopy without further purification. Many of these also conform to Reagent or Reagent ACS standards.
HPLC Grade	This is a designation for organic solvents which have been specifically prepared for use as HPLC mobile phases.

Second, the integrity of inventoried chemicals, or chemicals that have been on the shelf for a period of time, can be determined. This determination of integrity, or shelf-life, can involve proper labeling at the manufacturing plant, various reference sources, such as the *Merck Index*, or various laboratory tests to determine purity.

Third, samples and solutions gathered or prepared by laboratory personnel must also be properly labeled at the time of sampling or preparation. This means that the label must include the name of the sample or the chemical(s) dissolved and the date the sample was gathered or solution prepared. In addition, the laboratory should maintain a notebook of all samplings and preparations giving details of each sampling or preparation and the references which called for the particular chemicals or techniques used. This would allow tracking the reagent or sample in the event an error is suspected.

We made the point in Chapter 1 that "an analysis is only as good as the sample" used. We can now extend this statement and say that an analysis is only as good as the sample *and the reagents* that are used.

2.5 LABORATORY SAFETY

The analytical chemistry laboratory is a very safe place to work. The reason for this is that the dangers associated with contact with hazardous chemicals, flames, etc. are very well documented. As a result, laboratories are constructed and procedures are carried with these dangers in mind. Hazardous chemical fumes, for example, are vented into the outdoor atmosphere by fume hoods. Safety showers for diluting spills of concentrated acids on clothing are now commonplace. Eye wash stations are strategically located for the immediate washing of one's eyes in the event of accidental contact of a hazardous chemical with the eyes. Fire blankets, extinguishers, and sprinkler systems are also located in and around analytical laboratories for immediately extinguishing flames and fires. A variety of safety gear, such as safety glasses, aprons, and shields, is available. There is never a good excuse for personal injury in a well-equipped laboratory where well-informed analysts are working.

While the pieces of equipment mentioned above are now commonplace, it remains for the analysts to be well informed of potential dangers and of appropriate safety measures. Chemistry analysts should be well versed in the procedures used in the event of an accident at his/her worksite. These include (1) total familiarity with the safety features of the site, meaning location of the eye wash stations, safety showers, fire extinguishers; (2) total familiarity with the employer's laboratory safety manuals, and if one does not exist, it should be the analyst's responsibility to urge management to write one; and (3) knowledge of various reference books devoted to or at least emphasizing safety.[*]

An important development with regard to safety is the requirement of chemical manufacturers to provide Material Safety Data Sheets (MSDS) with all chemicals sold. The MSDS is a piece of paper that provides essential safety information concerning the chemical, such as reactivity properties, fire hazards, identification information, what to do in case of a spill, what special equipment is required for handling, and much more. MSDSs should be stored in an easily accessible location known to all laboratory workers.

---

[*] See, for example, Shugar, G.J. and Ballinger, J.T., *Chemical Technicians' Ready Reference Handbook*, 3rd ed., McGraw-Hill, New York, 1990, and references contained therein.

Laboratory work can be very important work indeed, but nothing is so important that a person's safety is jeopardized.

CHAPTER 3

SOLUTION PREPARATION

3.1 INTRODUCTION

One activity that is common to all analytical laboratories is solution preparation. Solutions must be prepared no matter what analytical method is employed or sample analyzed. This includes all wet chemical methods, as well as instrumental methods. Standard solutions must be prepared in order to generate a standard curve in instrumental analysis. Solutions must be prepared and standardized in order to perform titrations and to calculate the results of titrations. Buffer solutions must be prepared in order to provide a constant pH in solutions that require it. Solutions of reagents required for the chemistry of a given system are often needed for addition to the system at an appropriate time. Solutions are even often prepared for safety measures in the event of the need to neutralize a chemical spill. The list seems to be endless.

The accuracy with which the chemist prepares a solution depends, of course, on the application; some require the highest degree of accuracy, while others require only minimal accuracy (Chapter 1). If it is required that a solution concentration be of high accuracy, such as for the titrant in a titrimetric method or for a solution used to help generate a standard curve in an instrumental method, and if a standardization procedure

(Chapter 4) is not planned, then its concentration must be known accurately directly through its preparation. This means that an analytical balance must be used to weigh the chemical (if the solute is a pure solid), an accurate pipet must be used to measure the volume of a solution of the chemical (if the chemical is already in solution), and a volumetric flask must be used to measure the total solution volume in order to know that just the correct amount of solvent has been added. (See Chapter 2 for a description of the volumetric flask and its use.) However, if the solution concentration need not be known accurately, then any balance, or any type of liquid transfer glassware if the solute is already in solution, would work, and a volumetric flask need not be used.

With these thoughts in mind, we proceed with a series of discussions giving specific instructions as to calculations and methods used in a variety of situations involving various concentration units.

3.2 DILUTION

One method of preparing solutions, as alluded to in the last section, is by dilution, i.e., the solute is already in solution, but a lower concentration of it is called for. Regardless of the units with which the concentration (or volume) is expressed, the general calculation and method of preparation is the same. The calculation uses the following formula:

$$C_B \times V_B = C_A \times V_A \qquad (3.1)$$

in which "C" refers to "concentration," "V" is "volume," "B" refers to "before dilution," and "A" to "after dilution." The concentration of the solution before dilution times its volume is equal to the concentration of the solution after dilution times its volume. The concentrations before and after dilution must be known, and the volume after dilution must be known. The volume before dilution is being calculated in order to know how much of the more concentrated solution to measure out and dilute. The units of concentration must be the same on both sides of the equation (e.g., percent, molarity, normality, etc.). The units of volume are also the same on both sides.

Example 1
How would you prepare 500 mL of a 2.0% NaCl solution from one that is 10%?

Solution 1

$$10 \times V_B = 2.0 \times 500$$

$$V_B = \frac{2.0 \times 500}{10}$$

$$= 100 \text{ mL}$$

100 mL of the 10% solution are measured into a 500 mL container, and water is added to the 500 mL line.

A common example of a dilution problem is encountered when wanting to dilute commercially available concentrated acids, such as concentrated HCl or concentrated H_2SO_4. When a particular molarity of the acid is called for, it is necessary to know the molarity of the concentrated reagent in order to use the dilution formula. Such molarities are rarely displayed on the labels of these acids. They are listed in Table 3.1.

There can be considerable confusion when a dilution procedure in the form of a "ratio" is called for. Apparently, many clinical and biological applications call for a 1:10 dilution or a 2:5 dilution, etc. There is a difference in using the word "dilution" as opposed to the word "ratio." A 1:10 *dilution* is meant to mean 1 part solute added to make 10 parts total solution. A 1:10 *ratio*, or other synonym, is meant to mean 1 part solute added to 10 parts of solvent.

Additional dilution problems, those in which the concentrations before and after dilution have different units, and others are variously cited in the next four sections.

3.3 PERCENT

The concept of percent is well known. For solution preparation in an analytical laboratory, however, this concept can be confusing. The kind of units used for the numbers can be variable. Not only is it common for a solution concentration to be expressed as a volume/volume percent (the numerator and denominator both have volume units) or a weight/weight percent (the numerator and denominator both have weight units), but it is also common that the numerator has weight units, and the

Table 3.1 Molarities of Some Common Commercially Available Concentrated Acid and Base Solutions

Acid or Base	Molarity
Acetic acid ($HC_2H_3O_2$)	17
Ammonium hydroxide (NH_4OH)	15
Hydrobromic acid (HBr)	9
Hydrochloric acid (HCl)	12
Hydrofluoric acid (HF)	26
Nitric acid (HNO_3)	16
Perchloric acid ($HClO_4$)	12
Phosphoric acid (H_3PO_4)	15
Sulfuric acid (H_2SO_4)	18

denominator has volume units (so-called weight/volume percent).

$$\text{volume/volume (v/v) percent} = \frac{\text{volume of solute}}{\text{volume of solution}} \times 100 \qquad (3.2)$$

$$\text{weight/weight (w/w) percent} = \frac{\text{weight of solute}}{\text{weight of solution}} \times 100 \qquad (3.3)$$

$$\text{weight/volume (w/v) percent} = \frac{\text{weight of solute}}{\text{volume of solution}} \times 100 \qquad (3.4)$$

Note that the denominator in each case is the quantity of *solution* and not solvent. Also, the units in Equations 3.2 and 3.3 can be any unit of weight or volume as long as they are the same in both numerator and denominator. The units in Equation 3.4 must be grams and milliliters, kilograms and liters, or milligrams and microliters, etc.

3.3.1 Volume/Volume (v/v)

A volume/volume percent is common when the solute is a liquid. The reason is simply that volumes of liquids are easier to measure than weights. They can be pipetted. The following formula is useful.

$$\text{volume to measure} = (\%\ \text{desired}/100) \times \text{volume desired} \qquad (3.5)$$

The solute is measured (pipetted, if accuracy is important) into the container (a volumetric flask, if accuracy is important), and the solvent is added such that the total volume is the volume of solution desired.

Example 2
How would you prepare 500 mL of a 15% (v/v) ethanol solution in water?
Solution 2

$$\text{volume to measure} = (15/100) \times 500$$

$$= 75 \text{ mL}$$

75 mL of ethanol are measured into the vessel to contain the solution. Water is then added to the 500 mL level.

3.3.2 Weight/Weight (w/w)

A weight/weight percent is common when the solute is a solid, since a solid is easily weighed, but may also be encountered when the solute is a pure liquid. If the solute is a pure liquid, its volume is more easily measured, and it would be beneficial to calculate the volume corresponding to the weight (if the density is known). In addition, the weight of the total solution desired must be known. Thus, in the final stage of preparation, the solvent must be added such that the total weight is measured and matched to the weight of solution desired. Alternatively, if the density of the final solution is known, its volume may be measured instead. The following formula is useful.

$$\text{weight to measure} = (\% \text{ desired}/100) \times \text{solution weight desired} \qquad (3.6)$$

Example 3
How would you prepare 500 g of a 12% (w/w) solution of NaCl?
Solution 3

$$\text{weight to measure} = (12/100) \times 500$$

$$= 60 \text{ g}$$

60 g of NaCl are weighed, and placed in a container capable of holding 500 g of solution. Solvent is added until the solution weighs 500 g.

If the desired volume of the solution is given rather than the weight, the weight of the solution can be determined using the known density.

$$\text{weight of solution} = \text{volume of solution} \times \text{density} \qquad (3.7)$$

Example 4
How would you prepare 500 mL of a 12% (w/w) solution of NaCl if the density of the solution is 1.05 g/mL?
Solution 4

$$\text{weight of solution} = 500 \times 1.05 = 525 \text{ g}$$

$$\text{weight to measure} = (12/100) \times 525 = 63 \text{ g}$$

63 g of NaCl are weighed and placed in the vessel to contain the solution. The solvent is added up to the 500 mL level.

If the density is not known, it may be acceptable to assume that the density of dilute water solutions is equal to 1 (the density of pure water is equal to 1 at 4°C) such that the solution weight required (in grams) is numerically equal to the volume required (in milliliters).

3.3.3 Weight/Volume (w/v)

The concept alluded to in the last paragraph of assuming that the density of a dilute water solution is equal to 1 likely gave rise to the weight/volume expression for solution concentration. It is a convenient concentration expression because the amount of solute to be weighed is a percentage of the total solution volume rather than total solution weight.

$$\text{weight to measure} = (\% \text{ desired}/100) \times \text{solution volume desired} \qquad (3.8)$$

Thus, the solute can be weighed, placed in the container, and water added

up to the solution volume desired without regard for the total weight. This means that the solution does not have to be weighed, nor do we need to make the possibly invalid assumption that the density is equal to 1.

Example 5
   How would you prepare 500 mL of a 5.0% (w/v) solution of NaCl?
Solution 5

$$\text{weight to measure} = (5.0/100) \times 500$$
$$= 25 \text{ g}$$

25 g of NaCl are weighed and placed in the container. Water is then added such the total solution volume is 500 mL.

One source of confusion is in the preparation of weight/volume solutions from commercial concentrated acids, such as HCl, HNO_3, etc., which have known weight/weight percent concentrations. The density (or specific gravity) of the concentrated acid is used in the calculation and can often be found on the label of the concentrated acid bottle along with the weight/weight percent. This may be considered a dilution problem (see section 3.2), but the two concentrations, before dilution and after dilution, must have the same unit in order to correctly apply the dilution formula. We must convert weight/weight percent to weight/volume percent.

$$\text{w/v\%} = \text{w/w\%} \times \text{specific gravity} \qquad (3.9)$$

We thus will have a weight/volume percent that we can plug into the dilution formula.

Example 6
   How would you prepare 500 mL of a 10.0% (w/v) solution of HCl from concentrated HCl, which is 37.0% (w/w)? The specific gravity of the HCl, from the label on the bottle, is 1.18.
Solution 6

Solution 6

$$C_B(w/v\%) = 37.0 \ (w/w\%) \times 1.18$$

$$= 43.7\% \ (w/v)$$

$$43.7\%(w/v) \times V_B = 10.0 \times 500$$

$$V_B = \frac{10.0 \times 500}{43.7}$$

$$V_B = 114 \ mL$$

114 mL of concentrated HCl are diluted to 500 mL with water.

The percent composition (w/w) and the densities of the common concentrated acids and bases are given in Table 3.2.

Frequently, it is unclear whether a certain percent concentration called for in a given method is weight/weight or weight/volume. The analyst may need to make an educated guess in these instances. Since, for dilute water solutions, the density is very nearly equal to 1, the easiest method, usually the weight/volume, is chosen if accuracy is not important.

3.4 MOLARITY AND FORMALITY

The number of moles of solute dissolved per liter of solution (molarity) is another common method of expressing concentration. It is thus very common for an analyst to need to prepare solutions of a particular molarity, both by dilution and by weighing a pure chemical. In this case, solutions are referred to as being, for example, 2.0 molar or 2.0 *M*. The "*M*" refers to "molar," and we say that the solution has a molarity of 2.0, i.e., 2.0 moles dissolved per liter of solution. It is important to recognize that it is the number of moles dissolved per liter of *solution* and not per liter of solvent.

In addition, the terms "formality," "F," and "formal" are often encountered. These terms are meant to precisely describe solutions of ionic salts since, on an atomic scale, such salts technically do not exist as molecules, but rather as ions in a three-dimensional array. This is also the reason

Table 3.2 The Densities and Percent Compositions of Some Common Commercially Available Concentrated Acids and Bases

Acid or Base	Density	% Composition (w/w)
Acetic acid ($HC_2H_3O_2$)	1.05	99.5
Ammonium hydroxide (NH_4OH)	0.90	58
Hydrobromic acid (HBr)	1.52	48
Hydrochloric acid (HCl)	1.18	37
Hydrofluoric acid (HF)	1.14	45
Nitric acid (HNO_3)	1.42	72
Perchloric acid ($HClO_4$)	1.67	70
Phosphoric acid (H_3PO_4)	1.69	85
Sulfuric acid (H_2SO_4)	1.84	96

that the term "formula weight" is used as opposed to "molecular weight." Thus, the *formula* weight is the weight represented by the formula of the compound, which is also what *molecular* weight represents for nonionic compounds. The two are often used interchangeably. In the following discussions, the terms "molecular weight," "molar," and "M" will be used exclusively for ionic and nonionic compounds alike.

To prepare a solution of a certain molarity when a pure chemical is to be measured, the following formula is useful.

$$\text{grams to measure} = L_D \times M_D \times MW_{SOL} \qquad (3.10)$$

in which L_D refers to liters desired, M_D to molarity desired, and MW_{SOL} to molecular weight of solute. The grams of chemical thus calculated is weighed out, placed in the container, and water added to dissolve and dilute to volume.

Example 7
How would you prepare 500 mL of a 0.20 M solution of NaOH from pure, solid NaOH?
Solution 7

$$\text{grams to measure} = 0.500 \times 0.20 \times 40.00$$

$$= 4.0 \text{ g}$$

The analyst would weigh 4.0 g of NaOH, place it in a 500 mL container, add water to dissolve the solid, and then dilute to volume.

If the solute is a liquid (somewhat rare), the grams calculated from Equation 3.10 can be converted to milliliters using the density of the liquid. In this way, the volume of the liquid can then be measured, rather than its weight, and it can then be pipetted into the container.

Sometimes it may be necessary to prepare a solution of a certain molarity by diluting another solution, the concentration of which is known in, for example, weight/volume percent. As noted earlier (Section 3.2), however, the dilution formula can only be applied when the two concentrations have the same units. In this case, the weight/volume percent can be converted to molarity and the volume to be diluted calculated in the usual way.

Example 8

How would you prepare 500 mL of a 0.20 *M* solution of NaOH from an NaOH solution that is 10% (w/v)?

Solution 8

$$10\% \,(\text{w/v}) \;=\; 10 \text{ g} / 0.10 \text{ L}$$

$$=\; 100 \text{ g/L}$$

$$\text{Molarity} \;=\; \frac{100}{40.00} = 2.5 \text{ M}$$

$$2.5 \times V_B \;=\; 0.20 \times 500$$

$$V_B \;=\; 40 \text{ mL}$$

40 mL of the 10% NaOH would be measured into a 500 mL container and diluted to volume.

3.5 NORMALITY

Another method of expressing solution concentration is normality. Normality is the number of equivalents per liter. It is the typical concentration unit when working within the system in which quantities of chemicals are expressed as equivalents and equivalent weights. This system is completely analogous to the system of moles and molecular weights. That means that in all calculations performed in which the mole and the molecular weight are used (including solution preparation), the equivalent and equivalent weight may also be used. Normality is the number of equivalents of solute dissolved per liter of solution, while molarity is the number of moles of solute dissolved per liter of solution.

3.5.1 Concept of the Equivalent

To understand normality, we obviously must understand the concept of the equivalent. An equivalent is almost always either the same as the mole or it is some fraction of a mole, such as half a mole, a third of a mole, etc. What that fraction is depends on the reaction in which the substance takes part. The equivalent of a substance is that part of a mole which reacts on a one-to-one basis with another substance. If one mole of one substance reacts with one mole of another substance, it may be more convenient to work in terms of moles and molarity rather than equivalents and normality. If, however, one mole of one substance reacts with, say, two moles of another substance, then, if you want to work with normality, solution preparation schemes and associated calculations must involve the equivalent, rather than the mole, and the equivalent weight rather than the molecular weight.

How does one determine what an equivalent is or what an equivalent weight is? The answer depends on the *kind* of reaction in which the substance of interest takes part. The method for determining the equivalent weights for acid-base reactions, for example, is different from the method for oxidation-reduction reactions. Let us consider the acid-base case first.

3.5.1a Acid/Base

The equivalent weight of an acid is the molecular weight divided by the number of hydrogens donated per molecule in the reaction. The equivalent weight of a base is the molecular weight divided by the number of hydrogens accepted per molecule in the reaction. (By the Bronsted-Lowry Theory, an acid is a substance that donates hydrogens in a reaction, while a base is a substance that accepts hydrogens.) Let us look at a few examples.

$$HCl + NaOH \rightarrow NaCl + H_2O \tag{3.11}$$

$$H_2SO_4 + 2NaOH \rightarrow Na_2SO_4 + 2H_2O \tag{3.12}$$

$$2HCl + Ba(OH)_2 \rightarrow BaCl_2 + 2H_2O \tag{3.13}$$

$$H_3PO_4 + 2NaOH \rightarrow Na_2HPO_4 + 2H_2O \tag{3.14}$$

The equivalent weights of both HCl and NaOH in Equation 3.11 are equal to their respective molecular weights. One hydrogen is donated in the case of HCl (it has only one per molecule), and one hydrogen is accepted in the case of NaOH (it has only one hydroxide per molecule to accept hydrogens). The same can be said for the HCl in Equation 3.13 and the NaOH in Equations 3.12 and 3.14. Remember, the definitions state that it is the hydrogens donated or accepted *per molecule*. The H_2SO_4, however, has two hydrogens to be donated per molecule, and indeed they both get donated, since two water molecules were formed and there is no longer any hydrogens remaining with the sulfate on the right side. The equivalent weight of sulfuric acid in this case is the molecular weight divided by 2. Similarly, the $Ba(OH)_2$ in Equation 3.13 has two hydroxides to accept hydrogens, and indeed they both do accept hydrogens, since two water molecules were formed and there are no hydroxides remaining with the barium. The equivalent weight is the molecular weight divided by 2.

The H_3PO_4 in Equation 3.14 has three hydrogens to be donated. However, there is still one hydrogen remaining with the phosphate on the right side and therefore only two hydrogens were donated. Thus the equivalent weight of the H_3PO_4 is the molecular weight divided by 2. This does not preclude the possibility of H_3PO_4 donating either one or three hydrogens in some other reaction. The equivalent weight *depends on the reaction involved*.

Some general statements can be made concerning the above discussion. Sulfuric acid almost never donates just one of its hydrogens. Thus, its equivalent weight is almost always the molecular weight divided by 2. Hydroxides, such as the $Ba(OH)_2$, almost always donate all the OH groups that they have. Thus, the equivalent weight of a hydroxide is always the molecular weight divided by the number of hydroxides that it has per molecule.

3.5.1b Redox

With oxidizing agents or reducing agents, the equivalent weight is the molecular weight divided by the number of electrons involved in the balanced half-reaction *per molecule*. Thus again, we must know the reaction involved and also be able to isolate and balance the half-reactions both for the elements involved and for charge, since the charge balance gives the number of electrons. Let us again look at a few examples. First,

$$Fe^{+2} + MnO_4^- \rightarrow Mn^{+2} + Fe^{+3} \tag{3.15}$$

The balanced half-reactions for this reaction are

$$Fe^{+2} \rightarrow Fe^{+3} + 1e^- \tag{3.16}$$

and

$$5e^- + 8H^+ + MnO_4^- \rightarrow Mn^{+2} + 4H_2O \tag{3.17}$$

The equivalent weight of the iron compound used as the source of the Fe, such as $FeCl_2$, $Fe(NO_3)_2$, etc., is the molecular weight divided by 1 (one electron given up per molecule in Equation 3.16). The equivalent weight of the permanganate ($KMnO_4$, $NaMnO_4$, etc.) is the molecular weight divided by 5 (five electrons taken on per molecule in Equation 3.17).

Another example is the iodide in the iodide/iodine chemistry.

$$I^- + Cr_2O_7^{-2} \rightarrow I_2 + Cr^{+3} \tag{3.18}$$

The half reactions involved are

$$2I^- \rightarrow I_2 + 2e^- \tag{3.19}$$

and

$$6e^- + 14H^+ + Cr_2O_7^{-2} \rightarrow 2Cr^{+3} + 7H_2O \tag{3.20}$$

The equivalent weight of potassium iodide, or whatever the source of the iodide, is the molecular weight divided by 1 (remember — "per molecule"), while the equivalent weight of potassium dichromate, for example, is the molecular weight divided by 6.

3.5.1c Precipitation/Complexation

For precipitation and complexation reactions, the equivalent weight of the substance furnishing the cation is the molecular weight divided by the total positive charge present per molecule due to the cation. For the substance furnishing the anion or ligand, it is that weight that reacts with one equivalent of the substance furnishing the cation. Here are some examples.

$$AgNO_3 + NaCl \rightarrow AgCl + NaNO_3 \tag{3.21}$$

$$BaCl_2 + Na_2SO_4 \rightarrow BaSO_4 + 2NaCl \tag{3.22}$$

$$3Cu(NO_3)_2 + 2K_3PO_4 \rightarrow Cu_3(PO_4)_2 + 6KNO_3 \tag{3.23}$$

$$Ca^{+2} + EDTA^{-4} \rightarrow Ca(EDTA)^{-2} \tag{3.24}$$

$$Cu^{+2} + 4NH_3 \rightarrow Cu(NH_3)_4^{+2} \tag{3.25}$$

The equivalent weights of both reactants in Equation 3.21 are the molecular weights divided by 1. The equivalent weights of both reactants in Equation 3.22 are the molecular weights divided by 2. In Equation 3.23, the equivalent weight of $Cu(NO_3)_2$ is the molecular weight divided by 2. The equivalent weight of the K_3PO_4 is the molecular weight divided by 3. In Equation 3.24, the equivalent weights of the calcium compound and the EDTA are the molecular weights divided by 2. In Equation 3.25, the equivalent weight of the copper compound is the molecular weight divided by 2. The equivalent weight of the ammonia is the molecular weight multiplied by 2.

When there is no equation given, the equivalent weight is often determined ambiguously according to the above definition. If an analyst is asked to prepare 500 mL of a 0.10 N Na_2SO_4 solution, for example, with no equation to guide them, they would likely decide ambiguously that the equivalent weight is the molecular weight divided by 2, since the total charge due to the sodium is +2.

The most important complexation reaction is the reaction of metal cations with ethylenediaminetetraacetic acid (EDTA) (see Chapter 4) as in Equation 3.24 above. Since this is always a one-to-one reaction in terms of moles, the analyst almost always works with moles and molarity (or titer — see Chapter 4) in this application.

3.5.2 Solution Preparation

To prepare a solution of a certain normality (N) when a pure chemical is to be measured, the following formula is useful.

$$\text{grams to measure} = L_D \times N_D \times EW_{SOL} \tag{3.26}$$

in which L_D is the liters desired, N_D is the normality desired, and EW_{SOL} is the equivalent weight of the solute.

To prepare a solution of a certain normality when a solution of a certain greater normality is to be diluted, the usual dilution equation is useful (Equation 3.1). However, if only the molarity of the solution to be diluted is known, it is first necessary to convert to normality.

$$N_B = M_B \times \text{the number of equivalents per mole} \tag{3.27}$$

in which N_B is the normality before dilution, and M_B is the molarity before dilution. The number of equivalents per mole is, of course, the number by which the molecular weight is divided to obtain the equivalent weight.

Example 9
How would you prepare 500 mL of a 0.20 N solution of KH_2PO_4 (a pure, solid chemical) if it is to be used as in the following equation?

$$KH_2PO_4 + 2KOH \rightarrow K_3PO_4 + 2H_2O$$

Solution 9

$$\text{grams to measure} = 0.500 \times 0.20 \times \frac{136.09}{2}$$

$$= 6.8 \text{ g}$$

6.8 g of KH_2PO_4 are measured into a 500 mL container. Water is added to dissolve and dilute to volume.

Example 10

How would you prepare 500 mL of a 0.20 N solution of H_2SO_4 from a solution of concentrated H_2SO_4 which is 18 M?

Solution 10

$$N_B = 18 \times 2$$

$$= 36$$

$$C_B \times V_B = C_A \times V_A$$

$$36 \times V_B = 0.20 \times 500$$

$$V_B = 2.8 \text{ mL}$$

2.8 mL are measured into a 500 mL container and diluted to volume.

3.6 PARTS PER MILLION/BILLION

Solution concentration is often expressed in parts per million (ppm) or parts per billion (ppb). For such solutions, parts per million is most often assumed to mean milligrams of the indicated solute per liter of solution, and parts per billion to mean micrograms of such solute per liter of solution. Technically, this should be milligrams per kilogram, or micrograms per kilogram, but, as stated previously in the percent dis-

micrograms per kilogram, but, as stated previously in the percent discussion, since the density of water is nearly 1 at room temperature, the liter and the kilogram for such dilute solutions represent the same quantity. The grams to be measured (when the solute is a pure solid or liquid) are calculated as follows.

$$\text{grams to measure} = \frac{L_D \times ppm_D}{1000} \tag{3.28}$$

and

$$\text{grams to measure} = \frac{L_D \times ppb_D}{10^6} \tag{3.29}$$

in which L_D is the liters desired, ppm_D and ppb_D represent the concentrations desired, and 1000 and 10^6 are the conversion factors which convert milligrams and micrograms, respectively, to grams.

It may be more convenient or even required at times to weigh a quantity of a chemical that "contains" the solute rather than the pure solute. For example, it is more convenient to weigh the appropriate quantity of sodium chloride, rather than pure sodium, when preparing a solution of a certain parts per million sodium. Likewise, when preparing a solution of a certain parts per million nitrate, it would be necessary to weigh a quantity of a nitrate salt, such as potassium nitrate. In these cases, a "gravimetric factor," which would convert the weight of the solute to the weight of the substance actually weighed, would be required as part of the calculation.

$$\text{grams to measure} = \frac{L_D \times ppm_D \times \text{gravimetric factor}}{1000} \tag{3.30}$$

and

$$\text{grams to measure} = \frac{L_D \times ppb_D \times \text{gravimetric factor}}{10^6} \tag{3.31}$$

With both parts per million and parts per billion, the quantity weighed is often an extremely small quantity — frequently too small to be weighed accurately. If this is the case, a more concentrated solution should be prepared and then diluted. This would obviously involve a dilution

Also, for many solutions of this type, such as solutions of metals to be prepared for atomic absorption analysis (Chapter 7), the more concentrated solutions may be available commercially. With such solutions, then, only a dilution is required, but it may be necessary to convert from parts per billion to parts per million in order to have both concentration terms in the dilution equation of the same unit. In order to do this, the following equation is useful.

$$ppb = ppm \times 1000 \qquad (3.32)$$

Some examples follow.

Example 11
How would you prepare 500 mL of a 25.0 ppm copper solution using pure copper metal as the solute?
Solution 11

$$\text{grams to measure} = \frac{25 \times 0.500}{1000} = 0.0125 \text{ g}$$

0.0125 g of copper metal is weighed into a 500 mL flask, a nitric acid solution is added to dissolve the metal, and water is added to the 500 mL mark.

Example 12
How would you prepare 500 mL of a 100 ppb sodium solution using sodium chloride as the solute?
Solution 12
The gravimetric factor for converting the weight of sodium to sodium chloride is the molecular weight of sodium chloride divided by the atomic weight of sodium, which is 2.542.

$$\text{grams to measure} = \frac{100 \times 0.500 \times 2.542}{10^6} = 0.0001271 \text{ g}$$

Given such a small quantity of grams calculated, it would be expedient to prepare a much more concentrated solution (say 100 ppm) and dilute. So 100 ppm is 1000 times more concentrated, and thus would require 1000 times more grams, or 0.1271 g. Equations 3.1 and 3.32 would then prove useful.

$$ppb \quad = 100 \times 1000 = 100,000$$

$$100,000 \times V_B \quad = 100 \times 500$$

$$V_B \quad = 0.500 \text{ mL}$$

0.500 mL of the 100 ppm solution would be measured into a 500 mL flask and diluted to volume.

3.7 BUFFER SOLUTIONS

Buffer solutions, or solutions which resist changes in pH even when a strong acid or base is added, are almost always composed of a weak acid or weak base and the salt of this weak acid or base. The reason for the resistance to pH change is that the weak acid or weak base ionization equilibrium shifts in these solutions such that the H^+ or OH^- added are consumed, thus resulting in no net pH change.

Although commercially prepared buffer solutions are available, these are most often utilized solely for pH meter calibration and not for adjusting or maintaining a chemical reaction system at a given pH. It is not surprising, therefore, that the analyst often needs to prepare his/her own solutions for this purpose. It then becomes a question of what proportions of the acid, or base, and its salt should be mixed to give the desired pH.

The answer is in the expression for the ionization constant, K_a or K_b, where the ratio of the salt concentration to the acid concentration is found. In the case of a weak acid,

$$HA \rightleftharpoons H^+A^- \tag{3.33}$$

$$K_a = \frac{[H^+][A^-]}{[HA]} \tag{3.34}$$

and in the case of a weak base,

$$B + H_2O \rightleftharpoons BH^+ + OH^- \tag{3.35}$$

$$K_b = \frac{[BH^+][OH^-]}{[B]} \qquad (3.36)$$

Knowing the value of K_a or K_b for a given weak acid or base and knowing the desired pH value, one can calculate the ratio of salt concentration to acid (or base) concentration that will produce the given pH. Rearranging Equation 3.34, for example, would show the method for calculating this ratio in the case of a weak acid and its salt.

$$\frac{K_a}{[H^+]} = \frac{[A^-]}{[HA]} \qquad (3.37)$$

This is one form of the Henderson-Hasselbalch equation for dealing with buffer solutions. It says that one would simply divide the K_a by the $[H^+]$ to obtain the required ratio. It should be stressed that since the K_a, or K_b, enters into the calculation, how weak the acid or base is dictates what is a workable pH range for that acid or base. Table 3.3 gives commonly used examples of acid/salt and base/salt combinations and for what pH range each is useful.

It should also be stressed that the pH value of an actual buffer solution, prepared by mixing quantities of the weak acid or base and its salt based on the calculated ratio, will likely be different from what is expected. The reason for this is the use of approximations in the calculations, such as the use of molar concentration rather than activity, and the fact that the values for the un-ionized acid and base concentrations in the denominators in Equations 3.34 and 3.36 are approximations. A better method of preparing a buffer solution would be to prepare a solution of the salt (the concentration is not important) and then add a solution of a strong acid (or base, if the salt is a salt of a weak base) until the pH, as measured by a pH meter, is the desired pH. The strong acid (or base), in combination with the salt, creates the equilibrium required for the buffering action. For example, to prepare a pH = 9 buffer solution, one would prepare a solution of ammonium chloride (refer to Table 3.3) and then add a solution of sodium hydroxide while stirring and monitoring the pH with a pH meter. The preparation is complete when the pH reaches 9. The equilibrium created would be

$$NH_4OH \rightleftharpoons NH_4^+ + OH^-$$

Table 3.3   **Commonly Used Examples of Acid/Salt and Base/Salt Combinations and Corresponding pH Ranges**

Combination	pH Range
Trichloroacetic acid + sodium trichloroacetate	1.8–3.8
Acetic acid + sodium acetate	3.7–5.7
Sodium dihydrogen phosphate + sodium monohydrogen phosphate	6.1–8.1
Ammonium hydroxide + ammonium chloride	8.3–10.3

Table 3.4   Recipes for Some of the More Popular Buffer Solutions

pH = 4.0 phthalate buffer	Dissolve 10.12 g of dried potassium hydrogen phthalate (KHP) in 1 L of solution
pH = 6.9 phosphate buffer	Dissolve 3.39 g of dried potassium dihydrogen phosphate and 3.53 g of dried sodium monohydrogen phosphate in 1 L of solution
pH = 10.0 ammonia buffer	Dissolve 70.0 g of dried ammonium chloride and 570 mL of concentrated ammonium hydroxide in 1L of solution

and the solution would be a solution containing a weak base (NH_4OH) and its salt (NH_4Cl).

Recipes for standard buffer solutions can be useful, however. Table 3.4 gives specific directions for preparing some popular buffer solutions.

CHAPTER 4

WET METHODS AND APPLICATIONS

4.1 INTRODUCTION

There are two classical methods of chemical analysis that are referred to as "wet" methods. These are "gravimetric" analysis and "titrimetric" (or "volumetric") analysis. They are called wet methods because they rarely make use of any electronic instrumentation beyond the analytical balance. They employ physical separation schemes and/or chemical reactions and classical reaction stoichiometry as the sole basis for arriving at the results. Laboratories referred to as "wet laboratories" are those in which analyses and preparations involving titrating, weighing, extracting, etc. are performed. Such procedures can be sample pretreatment procedures, such as extractions, that are performed in advance of instrumental analyses, as well as the self-contained gravimetric and titrimetric procedures.

4.2 GRAVIMETRIC ANALYSIS

Gravimetric analysis is the classical wet method characterized by the fact that only one kind of measurement, that of weight, is made on the sample, its constituents, and reaction products. Only measurements of weight are thus used in the calculation.

Gravimetric procedures always involve the separation of the analyte constituent from the sample so that it can be isolated and weighed. This separation can be a physical separation, such as through solubility or volatilization, or it can be a chemical separation, i.e., by chemical reaction, so as to form a precipitate that can be weighed. Physical separations for gravimetric analysis are often conceptually simple. An example would be the analysis of a wastewater sample for suspended solids and total solids. The procedures for such analytes are purely gravimetric and involve simple physical separation schemes. In the case of total solids, a particular volume of a wastewater sample is added to a preweighed evaporating dish, the water is evaporated in an oven, and the dish is weighed again. The difference in the two weights is the total solids in the sample. For suspended solids, a preweighed filtering crucible is used to filter the undissolved solids from a given volume. After drying, this crucible is weighed again, and the difference in the two weights is the suspended solid content of that sample. Similarly, moisture or any volatile component in various types of samples can be determined by evaporation and weighing the container before and after.

Classical gravimetric analysis, which is now mostly obsolete, involved the chemical separation scheme referred to above. An example would be the analysis of a powdered sample for sulfate content. In this classical example, the sulfate is separated by reacting it with a selective reagent, such as a solution of a barium salt, such that an insoluble precipitate, barium sulfate in this case, is formed. The precipitate is then filtered, dried, often "ignited," and weighed, with the analyte calculated through classical reaction stoichiometry. Since this "chemical" separation scheme is now mostly obsolete, this method will not be considered further in this text.

4.3 TITRIMETRIC ANALYSIS

The other classical wet chemical method, titrimetric analysis, unlike gravimetric analysis, involves the accurate measurement of volume, as well as weight, and is *not* obsolete. A technique known as a "titration" is at the heart of the method. In a titration, a buret (Chapter 2) is used

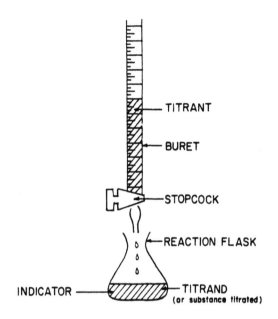

FIGURE 4.1   The equipment and terminology of a titration experiment. (From Kenkel, J., *Analytical Chemistry for Technicians*, Lewis Publishers, Inc., Chelsea, MI, 1988. With permission.)

to deliver a particular volume of a solution to a reaction flask. The substance held in the buret is called the "titrant." The substance originally held in the reaction flask is called the "titrand" or "substance titrated." In addition, a substance known as an "indicator" is often added to the reaction flask (see Figure 4.1). Typically, the titrant is added from the buret for the purpose of quantitatively reacting with the substance titrated. When the last bit of substance titrated is reacted with the added titrant (the so-called "equivalence point"), a change, such as a color change, is observed in the reaction flask. This change is usually caused by the chemistry of the indicator occurring at the equivalence point and is used to indicate when the addition of the titrant should be stopped. The analyst must strive to add a minute amount of titrant when near the equivalence point (so as to precisely arrive at the equivalence point) so that the precise amount of titrant required is then what is added and measured.*

---

*   While indicators are commonly used to detect equivalence points, there are other techniques which are described in later chapters, namely potentiometry and amperometry. The Karl Fischer method for water is a method that utilizes potentiometry. It will be discussed later in this chapter.

While the equivalence point is the exact point at which the last bit of substance titrated is reacted, the indicator usually does not respond precisely at that moment. Although this is not usually considered a problem, there is a distinction made between the point at which the reaction is complete and the point at which the reaction is complete *as indicated by the indicator*. The former is the equivalence point and the latter is the "end point."

A titration experiment usually takes place for two reasons: (1) for the standardization of a titrant and (2) for the quantitative determination of a constituent (analyte) in a sample dissolved and placed in the reaction flask. In either case, the volume added from the buret must be accurately read and recorded, and the quantity of substance titrated originally placed in the reaction flask must be accurately known. Let us now discuss each of these types of experiments.

4.3.1 Standardization

A "standard solution" is a solution which has a known concentration, meaning that it is known to the accuracy required for a given experiment. "Standardization" refers to an experiment in which the concentration of a solution is determined to the desired accuracy. To "standardize" a solution means to determine the concentration of a solution via a standardization experiment. A "primary standard" is the substance to which a solution is compared which allows standardization to take place.

This last definition warrants some additional comment. Obviously, the quality of the primary standard substance is ultimately the basis for a successful standardization. This means that it must meet some special requirements with respect to purity, etc., and these are enumerated in the following:

1. It must be 100% pure, or at least its purity must be known.
2. If it is impure, the impurity must be inert.
3. It should be stable at drying oven temperatures.
4. It should not be hygroscopic; it should not absorb water when exposed to laboratory air.
5. The reaction in which it takes part must be quantitative and preferably fast.
6. A high molecular weight is desirable.

Most substances used as a primary standard can be purchased as primary standard grade (Chapter 2), and this is appropriate and sufficient for a standardization experiment.

There are two general approaches to a standardization. One is to weigh the primary standard directly into the reaction flask. This approach could be called a standardization with a primary standard. The other approach is to measure a volume of a solution into the reaction flask; a solution that already has a known concentration (standard solution) or at least a concentration that will become known in subsequent experiments. This approach can be called standardization with a standard solution. In this latter approach, the concentration of the standard solution used can be known directly through its preparation, if the solute is a primary standard, or it can be determined through an independent standardization experiment. In any case, the calculation involved is based on the fact that the equivalents of titrant added at the end point are equal to the equivalents of substance titrated originally in the flask (see Chapter 3 for the definition of the "equivalent").

$$\text{equivalents of titrant (T)} = \text{equivalents of substance titrated (ST)} \quad (4.1)$$

or

$$\text{equiv.}_T = \text{equiv.}_{ST} \quad (4.2)$$

If a primary standard was weighed into the reaction flask, then Equation 4.2 becomes

$$L_T \times N_T = \frac{\text{grams}_{ST}}{\text{equiv. wt.}_{ST}} \quad (4.3)$$

in which L_T is the liters of titrant (the buret reading converted to liters); N_T is the normality of the titrant (the concentration — see Chapter 3); grams_{ST} is the weight of the substance titrated, the primary standard; and equiv. wt._{ST} is the equivalent weight of this substance titrated. In the actual experiment, N_T is the only unknown in Equation 4.3, and thus it can be calculated.

If a standard solution was pipetted into the reaction flask, then Equation 4.2 becomes:

$$L_T \times N_T = L_{ST} \times N_{ST} \tag{4.4}$$

in which L_{ST} is the volume pipetted, and N_{ST} is the normality of the solution pipetted. Here again, N_T is the only unknown, and thus it can be calculated. If a primary standard was used to prepare the solution to be pipetted into the reaction flask, N_{ST} can be calculated as follows:

$$N_{ST} = \frac{\text{grams}_{PS}/\text{equiv. wt.}_{PS}}{\text{liters prepared}} \tag{4.5}$$

in which "PS" refers to "primary standard." A volumetric flask (Chapter 2) is required for this, since the "liters prepared" needs to be known accurately.

One further comment concerning Equation 4.4 and this experiment in general is that the solution pipetted into the reaction flask, rather than the titrant, may be the solution to be standardized. In that case, N_{ST} in Equation 4.4 is the unknown concentration to be calculated and N_T is known.

It should be mentioned that a standardization experiment can be avoided completely if the titrant solute is a primary standard material, and the concentration can thus be known through its preparation. This would involve a procedure identical to that described above for the standard solution used in "standardization with a solution" — weighing the solute accurately into a volumetric flask and calculating the normality as in Equation 4.5.

Finally, a concentration can be represented as "titer." The titer of a solution is the weight of a substance that reacts with 1 mL of the solution, such as a titrant. The calculation here involves dividing the grams of primary standard present in the reaction flask by the milliliters of titrant needed to titrate it. This would be the titer of the titrant and a measure of its strength or concentration. See the determination of water hardness and the Karl Fischer method for water in organic solvents in Section 4.4 for example applications of this concept.

4.3.2 Titration of Unknowns

The quantitative determination of an analyte in a solution is the other use for a titration experiment. In such an experiment, in contrast to the standardization experiment, the titrant concentration is known, either directly through its preparation or because it was standardized, while the quantity of substance titrated, the analyte, is unknown. In such an experiment, the sample is measured, pipetted or weighed, into the reaction flask and titrated with the titrant to a suitable end point.

The calculation used depends on what result is desired. It is common to calculate the percent of an analyte in a sample that was weighed into the reaction flask, in which case the following derivation is useful.

$$\% \text{ analyte} = \frac{\text{analyte wt.}}{\text{sample wt.}} \times 100 \tag{4.6}$$

The analyte weight can be calculated from the titrant's normality and volume and the equivalent weight of the analyte:

$$\text{analyte wt.} = L_T \times N_T \times \text{equiv. wt.}_{\text{analyte}} \tag{4.7}$$

Combining Equations 4.6 and 4.7, we obtain the equation for the percent of an analyte:

$$\% \text{ analyte} = \frac{L_T \times N_T \times \text{equiv. wt.}_{\text{analyte}}}{\text{sample wt.}} \times 100 \tag{4.8}$$

Sometimes parts per million of analyte may be calculated, particularly if the sample is a liquid. The water hardness titration discussed in Section 4.4 is an example of such a titration.

4.4 EXAMPLES OF REAL-WORLD TITRIMETRIC ANALYSIS

4.4.1 The Kjeldahl Method

A titrimetric method that has been used for many years for the determination of nitrogen and/or protein in a sample is the Kjeldahl method. Examples of samples include grain, protein supplements for animals feed, fertilizers, and food products. It is a method that may use a special technique called a "back titration." We will now describe this technique in detail.

The method consists of three parts: (1) the digestion, (2) the distillation, and (3) the titration. The digestion step is in essence the dissolving step. The sample is weighed and placed in a Kjeldahl flask, which is a round bottom flask with a long neck similar in appearance to a volumetric flask, except for the round bottom and the lack of a calibration line. A fairly small volume of concentrated sulfuric acid along with a quantity of K_2SO_4 (to raise the boiling of the sulfuric acid) and a catalyst (typically an amount of $CuSO_4$, selenium, or a selenium compound) is added, and the flask is placed in a heating mantle and heated. The sulfuric acid boils and sample digests for a period of time until it is evident that the sample is dissolved and a clear solution is contained in the flask. The digestion must be carried out in a fume hood, since thick SO_3 fumes evolve from the flask until the sample is dissolved. At this point, the contents of the flask is diluted with water, and an amount of fairly concentrated sodium hydroxide is added to neutralize the acid. Immediately upon neutralization, the nitrogen originally present in the sample is converted to ammonia. At this point the distillation step is begun.

A laboratory that runs Kjeldahl analyses routinely would likely have a special apparatus set up for the distillation. The essence of this apparatus is shown in Figure 4.2. A baffle is placed on the top of the Kjeldahl flask and subsequently connected to a condenser which in turn guides the distillate into a receiving flask as shown. The ammonia is thus distilled into the receiving flask. The receiving flask contains an acid for reaction with the ammonia.

The acid in the receiving flask can either be a dilute (perhaps 0.10 normal) standardized solution of a strong acid, such as sulfuric acid, or a solution of boric acid. If it is the former, it is an example of a "back titration." If it is the latter, it is an example of an indirect titration.

In the back titration method, an excess, but carefully measured, amount of the standardized acid is contained in the flask such that, after the ammonia bubbles through it and is consumed, an excess remains. The flask is then removed from the apparatus, and the excess acid is titrated with a standardized NaOH solution. The analyte in this procedure is the nitrogen (in the form of ammonia as it enters the flask), and thus the amount of acid consumed is the important measurement. The amount of acid consumed is the difference between the total amount present and the amount that was in excess. It is called a back titration because the amount of acid in the flask is in excess and in essence goes beyond the

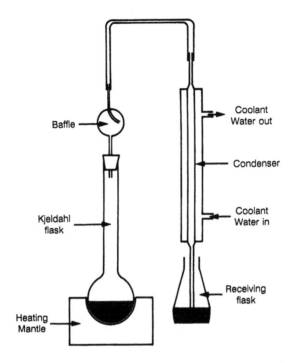

FIGURE 4.2   A typical apparatus for the Kjeldahl distillation step.

end point for the reaction with the ammonia. Thus, the analyst must come *back* to the end point with the sodium hydroxide. The calculation is.

$$\% \text{ N} = \frac{\left(L_T \times N_T - L_{BT} \times N_{BT}\right) \times 14.00}{\text{sample wt.}} \times 100 \qquad (4.9)$$

in which the titrant is the dilute sulfuric acid, and "BT" refers to "back titrant," the NaOH. The equivalent weight shown is the atomic weight of nitrogen. The percent protein may also be calculated, in which case the equivalent weight of the protein is substituted for the 14.00.

In the indirect method using boric acid, the ammonia reacts with the boric acid producing a partially neutralized salt of boric acid, $H_2BO_3^{-1}$, which can then be titrated with a standardized acid. The amount of standardized acid needed is proportional to the amount of ammonia that bubbled through. It is called an indirect method because the ammonia is determined by titration of the $H_2BO_3^{-1}$. In a direct titration, the analyte would be reacted directly with the titrant. Equation 4.8 is used for this as in the direct method.

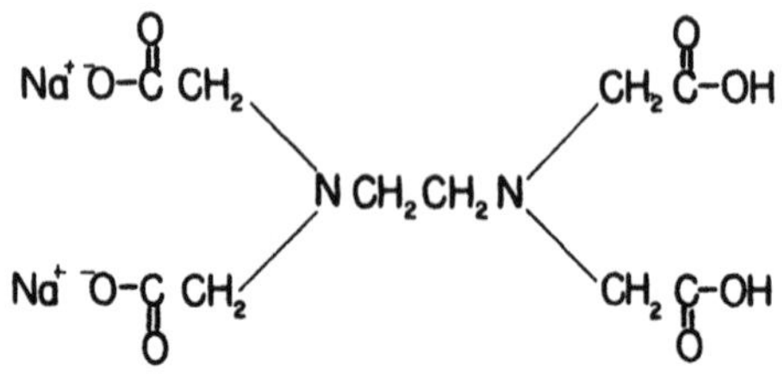

FIGURE 4.3   Disodium dihydrogen EDTA.

4.4.2 Water Hardness

The analysis of water samples for hardness is a common analysis in water and environmental laboratories. The usual procedure for this is a titrimetric procedure in which a solution of disodium dihydrogen ethylenediaminetetracetic acid dihydrate ($Na_2H_2EDTA \cdot 2H_2O$) is the titrant (see Figure 4.3). This compound furnishes a hexadentate ligand which reacts with the calcium, magnesium, and iron ions in hard water forming complex ions. The calcium, magnesium, and iron ions are the major contributors to water hardness. The EDTA is in the buret, and the water sample in the reaction flask.

A complex ion is a polyatomic charged aggregate consisting of a positively charged metal ion combined with a ligand. A ligand is an uncharged or negatively charged chemical species that reacts with a metal ion to form a complex ion. Ligands are classified according to the number of bonding sites in their structure available for bonding to the metal ion. A ligand with one site is "monodentate," one with two sites is "bidentate," etc. The EDTA species, in somewhat basic solution (pH = 10), is "hexadentate"; there are six bonding sites available for bonding to metal ions. The reaction with metal ions, however, is one-to-one. The calcium-EDTA complex ion is shown in Figure 4.4.

The usual indicator for this titration is eriochrome black T, an organic dye. Eriochrome black T is actually a ligand that also reacts with the metal ions, like EDTA. In the free uncombined form and at the basic pH, it imparts a sky blue color to the solution, but if it is a part of a complex ion with either calcium or magnesium ions, it is a wine red color. Thus, before adding any EDTA from a buret, the hard water sample, containing a pH = 10 buffer (such as an ammonia buffer — see Chapter 3) and several drops of the indicator solution, will be wine red. As the EDTA solution is added, the EDTA ligand reacts with the free metal ions and then

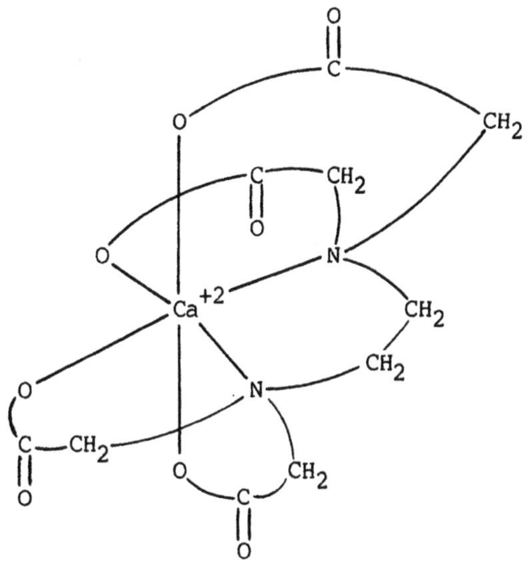

FIGURE 4.4   The calcium-EDTA complex ion.

actually reacts with the metal/indicator complex ion, displacing and complexing the metal ions and resulting in the free indicator ligand, which, as mentioned above, gives the solution a sky blue color. The color change, then, is the total conversion of the wine red color to the sky blue color, with every trace of red disappearing at the end point.

It is known that this color change is quite sharp when magnesium ions are present. In cases in which magnesium ions are not present in the water samples, the end point will not be sharp. Because of this, a small amount of magnesium chloride is added to the EDTA as it is prepared, and thus a sharp end point is assured. Also, other indicators, similar to eriochrome black T, are sometimes used; these are murexide and calmagite. Calmagite shows approximately the same end point and has an advantage in that its solutions are more stable with time. Eriochrome black T solutions need to be prepared fresh about every two weeks.

Water hardness is usually reported either as parts per million $CaCO_3$ or grains per gallon $CaCO_3$. The parts per million unit, as discussed in Chapter 3, represents the number of milligrams present per liter. Since it is easier to measure the volumes of liquids, as opposed to weight, a *volume* of water is measured (pipetted) into the reaction flask, rather than a weight, and thus it is logical and easy to calculate the milligram per liter, or parts per million.

$$\text{ppm CaCO}_3 = \frac{L_{EDTA} \times M_{EDTA} \times F.W. \ CaCO_3 \times 1000}{\text{liters of water}} \qquad (4.10)$$

The numerator is multiplied by 1000 to convert grams to milligrams. Also, multiplying by the formula weight of $CaCO_3$ in the numerator has the effect of reporting all metals that react with the EDTA as $CaCO_3$, a common practice in water laboratories.

Molarity and formula weight (F.W.) are used here since the reaction is one-to-one in terms of moles, and the moles of EDTA ($L \times M$) are thus converted directly to grams of $CaCO_3$ by multiplying by the formula weight of $CaCO_3$. Standardization experiments for the EDTA have as their goal, then, the molarity, rather than the normality of the solution. For this reason, Equations 4.3, 4.4, and 4.5 would involve molarity instead of normality, and formula weight instead of equivalent weight in this case. The primary standard for this experiment is typically a pure grade of $CaCO_3$.

The concentration of the EDTA for this titration is typically 0.01 M. Because of this, and since the F.W. of $CaCO_3$ is 100.09, if 100 mL of water is used, the results can be calculated by simply multiplying the buret reading (in milliliters) by 10.

Finally, the "titer" (see Section 4.3) of EDTA, or the number of grams of $CaCO_3$ that reacts with 1 mL of EDTA, can be used here.

$$\text{ppm CaCO}_3 = \frac{CaCO_3 \text{ titer of EDTA} \times \text{mL of EDTA} \times 1000}{\text{liters of water}} \qquad (4.11)$$

The parts per million unit can, of course, be converted to grains per gallon or vice versa. To convert parts per million to grains per gallon, we multiply the given parts per million by 0.05833. To convert grains per gallon to parts per million, we multiply the given grains per gallon by 17.14.

4.4.3 The Karl Fischer Titration

The Karl Fischer titration is a titration for the determination of small quantities of water in organic solvents or crystalline solids. It is an electroanalytical method (see Chapter 11) utilizing a pair of electrodes dipped into the reaction vessel. More precisely, it is a "bipotentiometric" technique in which a power source is connected across the electrodes so as to maintain a constant current between them. The power source may be as simple as a pH meter which has a pair of input terminals on it labeled "K-F" for Karl Fischer. In the modern laboratory, however, the Karl Fischer apparatus is typically a self-contained unit complete with titration cell, power source, automatic titrator (Chapter 12), reagent reservoir, and a computer, which includes digital and hardcopy readout and sometimes a keyboard with monitor. Since it is the water content that is being determined, moisture contamination must be scrupulously avoided. This means that the sample and titrant must be protected. Figure 4.5 shows a photograph of a modern Karl Fischer unit including all components referred to above (except the monitor). Drying tubes are shown protecting the titrant in the reservoir and the sample in the cell from atmospheric moisture.

The "titrant" for this method is the so-called Karl Fischer reagent. We have placed the word titrant in quotes here because the *sample* is actually added from the buret, not the reagent, although it is the reagent that must be standardized. The reagent is a mixture of chemicals, including iodine, pyridine, and SO_2, in a 1:10:3 ratio and usually dissolved in methanol solvent. Addition of the sample to this reagent begins a sequence of reactions which includes the consumption of water and iodine and the formation of iodide ion. In a solution which includes both iodine and iodide, the potential required to maintain the small constant current is small. However, at the moment that the last trace of iodine is consumed by the water from the added sample, the required potential is quite large, and there is a sudden shift to higher potential values, signaling the end point.

The typical standardization procedure involves the use of a sample of methanol with a known amount of dissolved water (a standard solution of water) prepared by adding a weight of water to a particular volume of water-free methanol solvent (or a volume of methanol solvent in which the water content is known) such that the milligrams per milliliter of water in the methanol is known. A carefully measured (pipetted) amount of the Karl Fischer reagent is placed in the cell, and the standard water sample is added to the end point. In this way the titer of the reagent, the milligram of water per milliliter of reagent, is determined.

$$\text{reagent titer} = \frac{\text{mg water/mL methanol solution} \times \text{mL methanol solution needed}}{\text{mL of reagent used}}$$

(4.12)

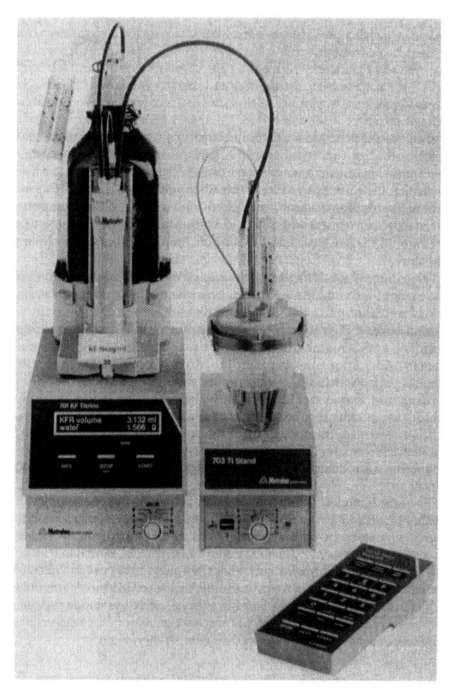

FIGURE 4.5   A photograph of a modern Karl Fischer titrator. (Photograph courtesy of Brinkman Instruments, Inc., Westbury, NY.)

The water content in an unknown sample can then be determined by multiplying the reagent titer by the milliliters of reagent used and dividing by the milliliters of sample needed.

$$\text{mg water/mL sample} = \frac{\text{reagent titer} \times \text{mL reagent used}}{\text{mL sample needed}} \qquad (4.13)$$

4.5 FINAL COMMENTS

In general, wet methods of analysis are used in the modern laboratory in situations in which (1) they are more convenient and/or accurate than instrumental methods and (2) analyte concentrations are so high that instrumental methods are not appropriate. The examples cited above are mostly a matter of the convenience and accuracy of the established method, although both the Kjeldahl and Karl Fischer methods, in their modern form, may be considered at least partially instrumental. Another example method based on convenience is "total alkalinity." Alkalinity is the capacity of dissolved solutes to neutralize an added strong acid. The strong acid is conveniently added from a buret, and the neutralization point (end point) is conveniently signaled with the use of an indicator or pH meter. This procedure is in common use in laboratories in which total alkalinity of water-based samples needs to be determined.

An example of an analyte concentration being too high for instrumental methods is in the analysis of plating solutions. Some, but not all, of the ingredients in such solutions are present at such high concentrations that the instrumental methods cannot be used without considerable dilution. It is not unusual to encounter titrimetric procedures in use in plating industry laboratories for such analytes as chloride, cyanide, sulfates, etc. The instrumental methods are usually most useful for low concentrations (parts per million level).

CHAPTER 5

INSTRUMENTAL METHODS — GENERAL DISCUSSION

5.1 INTRODUCTION

The remainder of this text is devoted to techniques and methods of chemical analysis involving electronic instrumentation. In Chapter 4, we made a distinction between wet methods and instrumental methods, saying that the wet methods known as gravimetric and titrimetric analysis are more classical as compared to instrumental techniques. They are classical techniques in the sense that they have adequately served the chemical analysis laboratory for many years, but have now largely been replaced by the streamlined and computerized instrumental methods, especially where there is a need to detect trace amounts of analytes and to analyze more complicated samples. There are certain aspects of instrumental analysis that are common to virtually all instrumental techniques. These aspects will be dealt with in detail in this chapter. Characteristics of the individual instruments and associated procedures will be covered in the chapters to follow.

5.2 INSTRUMENTAL DATA AND READOUT

5.2.1 Recorders

Most laboratory instruments measure some physical or chemical parameter of a solution of a sample and display the measurement as a proportional voltage on some type of readout device. This device can be a simple meter (like the display on a pH meter), digital or otherwise; a recorder, which may either be a strip-chart recorder or x-y recorder; a computer monitoring screen and/or a printer-plotter; or any combination of these. Let us first discuss the details of the two types of recorders and how they are used.

A recorder has one or more writing pens driven by a servomotor and records information on paper. The position of the pen on the paper represents the voltage or voltages fed into the recorder from the instrument, just as the position of the pointer on a meter represents such a voltage. The advantage of a recorder is that the voltage can be permanently recorded over a period of time, a capability which is of significant importance when the voltage output of an instrument changes with time, such as with gas chromatography (Chapter 9) or high performance liquid chromatography (Chapter 10), or when *two* parameters change with time, one as a function of the other, such as when recording a molecular absorption spectrum with a spectrophotometer (Chapter 6). This characteristic of voltage signals changing with time is quite common, and thus recorders are very commonly used in instrumental laboratories as readout devices.

The differences between the strip-chart recorder and the x-y recorder are noteworthy. The strip-chart recorder records the voltage output of an instrument strictly as a function of time. In other words, the pen records the voltage on the paper as the paper moves at a certain rate through the recorder. Thus, one axis of the resulting graph is time, and the other is the voltage level. Multiple pens in a strip-chart recorder can be used to record multiple voltage levels and how they change with time. The x-y recorder is a recording device which accepts two voltages from the instrument for recording with only one pen. One of the voltage levels determines the x-axis position of the pen, while the other determines the y-axis position. The paper does not move, but rather the pen moves over the paper as the two voltage levels change. Figure 5.1 shows representations

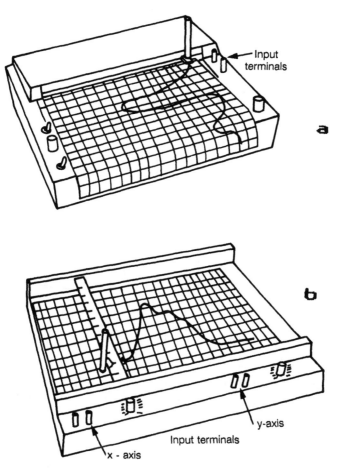

FIGURE 5.1   Drawings of (a) a strip-chart recorder (note only one set of input terminals for only one incoming signal), and (b) an x-y recorder (note two sets of input terminals for two incoming signals: one for the x-axis and one for the y-axis).

of (a) a strip-chart recorder and (b) an x-y recorder, while Figure 5.2 shows some sample recordings.

5.2.2 Instrument Readout and Concentration

Of course the goal of instrumental methods of analysis most of the time is the concentration of a constituent in an unknown sample. Thus, the

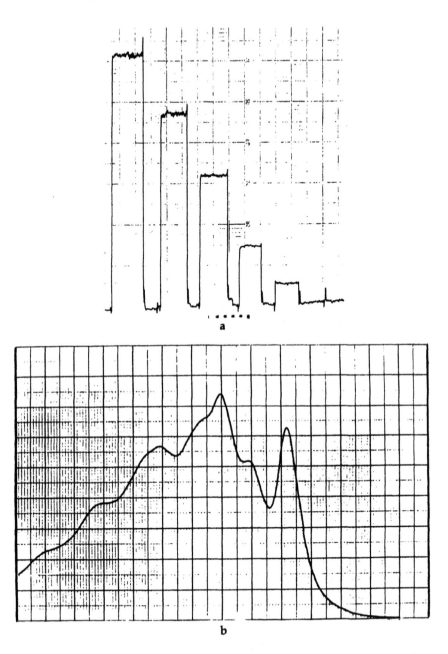

FIGURE 5.2   (a) The absorbances of a series of standard solutions recorded on a strip-chart recorder. (b) A molecular absorption spectrum recorded with an x-y recorder. (From Kenkel, J., *Analytical Chemistry for Technicians*, Lewis Publishers, Inc., Chelsea, MI, 1988. With permission.)

concentration of the desired constituent in the solution measured is to be determined from the readout resulting from that solution. As we have said, the readout (let us call it "R"), whatever it represents, is proportional to this concentration.

$$R = KC \qquad (5.1)$$

If the proportionality constant, K, is known, then the concentration can be calculated.

$$C = R/K \qquad (5.2)$$

Often, however, the proportionality constant is not known, and the experiment is not as simple as implied above. One alternate method would be to prepare a solution of the constituent so that the concentration is known (a standard solution), measure R, and then calculate K.

$$K = R_s/C_s \qquad (5.3)$$

In this equation, R_s is the readout for the standard, and C_s is the concentration of the standard. The value of K is then used to calculate C for the unknown, assuming, of course, that the parameters which contribute to the value of K for both solutions are identical at both concentrations.

$$C_u = R_u/K \qquad (5.4)$$

Here, R_u is the readout for the unknown, and C_u is the concentration of the unknown. Actually, the value of K need not be calculated at all, as is obvious from the following:

$$\frac{R_s = KC_s}{R_u = KC_u} \qquad (5.5)$$

or

$$\frac{R_s}{R_u} = \frac{C_s}{C_u} \qquad (5.6)$$

and

$$C_u = \frac{C_s R_u}{R_s} \qquad (5.7)$$

In other words, the value of C_u is related to that of C_s by the ratio of their instrument readings. Or, R_s is to R_u as C_s is to C_u.

This treatment requires that the parameters that contribute to the value of K not change between the two concentration levels, as indicated above. If the concentrations are very nearly the same, or if the relationship between the readout and concentration is of the same proportion at the different concentration levels, then this treatment is entirely valid. However, this may not be the case, and it is therefore useful to determine the linearity of R as a function of C over the concentration range in question. Such a linearity would verify the constancy of K. The procedure involves preparing a *series* of standard solutions and making a graph of R vs C to determine this linearity (see Figure 5.3a). If the results show linearity over the concentration range in question, the unknown concentration may be determined from the graph as shown in Figure 5.3b. Because of the uncertainty in knowing whether the proportionality constant is indeed constant over the concentration range studied, this series of standard solution method is commonplace in an instrumental analysis laboratory for virtually all quantitative instrumental procedures. Examples of this abound in this text for many spectrophotometric, chromatographic, and other techniques.

5.2.3 Method of Least Squares

The linearity of instrument readout vs concentration data must be established for best results. Random indeterminate errors during the solution preparation and during the measurement of R may cause the results to appear to deviate from linearity to some extent. In that case, a method must be adopted which will fit a straight line to the data as well as possible. It may happen that some (or even all) of the points may not fall exactly on the line because of these random errors, but a straight line must still be drawn, since the random errors are indeed random and cannot be compensated for directly. Thus, the best straight line possible is drawn through the points (see Figure 5.4).

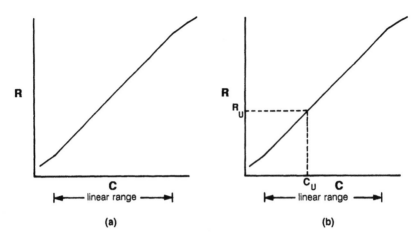

FIGURE 5.3   (a) A plot of an instrument readout, R, vs the concentration, C, of an analyte, showing linear relationship in the concentration range chosen. (b) The determination of an unknown's concentration from the graph in (a).

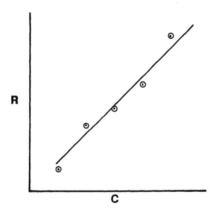

FIGURE 5.4   An example of a straight line fitted to a set of data.

"Eyeballing" the line through the points with a straightedge on the graph paper may easily result in significant error. Therefore, a procedure called the method of least squares (also called linear regression analysis) is best applied to the data, and this method results mathematically in the best straight line possible through a given set of data points. By this method, the best straight line fit is obtained when the sum of the squares of the individual y-axis value deviations (deviations between the plotted y values and the values on the proposed line) are at a minimum. This "proposed line" is actually calculated from the given data, a slope and y-intercept ("m" and "b," respectively, in the equation for a straight line, $y = mx + b$) are then obtained, and the deviations ($y_{point} - y_{line}$) for each given x are calculated. Finding the values of the slope and the y-intercept that minimize the sum of the squares of the deviations involves some complicated mathematics that is beyond the scope of this text. Computers and programmable calculators, however, handle this routinely in the modern laboratory, and the results are very important to a given analysis since the line that is determined is statistically the most correct line that can be drawn with the data obtained. The concentrations of unknown samples are also readily obtainable on the calculator or computer since the equation of the straight line, including the slope and y-intercept, are known as a result of the least squares procedure. Other statistically important parameters are readily obtainable as well, including the "correlation coefficient."

5.2.4 The Correlation Coefficient

The correlation coefficient is one measure of how well the straight line fits the analyst's data — how well a change in one variable correlates with a change in another. Laboratories can establish their own criteria as to what numerical value for such a coefficient is required for the accuracy desired. A correlation coefficient of exactly 1 indicates perfectly linear data. This, however, rarely occurs in practice. It occurs if all instrument readings increase by exactly the same factor as the concentration level increases, as in the following sample data:

R	C
4	2
8	4
12	6
24	12

Due to random errors, this data is more ideal rather than real. Data that approaches such linearity will show a correlation coefficient less than 1, but very nearly equal to 1. Numbers such as 0.9997 or 0.9996 are considered excellent and attainable correlation coefficients for many instrumental techniques. Good pipetting and weighing technique when preparing standards and well-maintained and calibrated instruments can minimize random errors and can produce excellent correlation coefficients and therefore accurate results. The analyst usually strives for *at least* two nines, or possibly three, in their correlation coefficients, depending on the particular instrumental method used. Again, these coefficients can be determined on programmable calculators and laboratory computers.

5.3 METHODS FOR QUANTITATIVE ANALYSIS

As we determined in the last section, the usual procedure for instrumental analysis is to plot the instrument readout for a series of standard solutions vs the concentration of those solutions, establish the linear range, and then, after determining an unknown readout, find its concentration by interpolation within this linear range. We now describe further details and several options that are useful when specific instrument and experiment designs dictate.

5.3.1 Series of Standard Solutions (or Serial Dilution) Method

When the instrument readout is totally free of matrix effects, sample loss effects, or any other such potential error-causing effects, the method described thus far works well. The usual procedure is to accurately prepare a stock standard solution of the analyte and then to prepare the series of standards by diluting this stock. Chapter 3 gives the specifics of various solution preparation schemes which may be employed here for both the stock and the series of standards, including preparing solutions by weighing a pure solid chemical and also by dilution. A possibility is to prepare the stock, dilute it to make the first standard, dilute the first to make the second, and the second to make the third, etc. This is called "serial dilution." In either case, the solution to be diluted is pipetted into a volumetric flask and diluted to the mark with either the pure solvent, often distilled water, or a solution determined to be an approximation of the sample matrix, if that is important, or having other additives, such as for pH adjustment. It should be mentioned that stock solutions of many analytes, especially metals (atomic absorption standards), are available commercially, and this eliminates the need to devote lab time to such preparation. This not only saves time, but also minimizes the possibility of error in this part of the procedure.

5.3.2 Internal Standard Method

In some instrumental procedures, the instrument readout can vary from standard to standard not only because of the concentration differences, but also because of uncontrollable experiment parameters. Such parameters may involve, for example, the problem of irreproducibility of sample injection volume in gas chromatography or nonlinear changes in solution viscosity as analyte concentration increases in flame AA. The solution to these problems is the Internal Standard Method.

In this method, all standards and the sample are spiked with a constant known amount of a substance to act as what is called an internal standard. The purpose of the internal standard is to serve as a reference point for instrument readout values for the analyte such that in a ratio of the readout for the analyte to the readout for the internal standard, the effect cancels out. Thus, if one were to plot this ratio vs the analyte concentration, one expects a linear relationship, and the problem goes away. Slight variations in gas chromatography (GC) injection volume, for example, are compensated for by the fact that the internal standard peak size and the analyte peak size (refer to Chapter 9) are both affected proportionally by such variations. A plot of the peak size ratio vs analyte concentration would be free of peak size variations due to irreproducible sample volume and thus would give accurate results.

Solution preparation here is identical with the Series of Standard Solutions Method, except that the constant known amount of internal standard is added (usually pipetted) to all standards and also to the unknown.

5.3.3 Method of Standard Additions

It is possible for the analyte readout in an instrumental procedure to be either suppressed or enhanced by some components of the sample. There is therefore a potential for error when these components are present in the sample but not in the standards. One solution to this problem would be to add the interfering substances to the standards as well so that the effect would be measured in both the standards and the sample. This procedure is called "matrix matching." The requirements of matrix matching are (1) that the interfering substances be identifiable and (2) that the concentrations of the interfering chemicals in the sample be known so that they can be matched in the preparation of the standards. The absence of either one or both of these would make matrix matching impossible. The answer is the Method of Standard Additions.

In this method, small amounts of a standard solution of (or pure) analyte are added to the sample, and the absorbance is measured after each addition. In this way, the interfering components need not be identified, and the sample matrix is always present at the same component concentrations as in the sample, with only perhaps a minor dilution.

The plotting procedure and the determination of the unknown concentration is altered somewhat, however. The plot is a graph of instrument readout vs concentration *added*. The first point to be plotted would be for 0 added (the sample readout), and the readout would increase (presumably linearly) for each addition of analyte. Extrapolation of the resulting line to zero readout (the x intercept), as shown in Figure 5.5, results in a length of x-axis on the negative side of zero added, which represents the concentration in the unknown as shown. In this method, one must presume that the plot is linear between the real zero and zero added, since the standards will not encompass that concentration region.

This method can be used in cases in which there is some sample preparation as well; for example, in cases in which lanthanum needs to be added in an atomic absorbance analysis for calcium. Once the pretreatment establishes the sample matrix, the standard additions can be performed and data graphed.

Since some sample may be consumed, such as in sample aspiration in AA or sample injection in GC, prior to making the second and subsequent additions of the analyte, the standard additions method could result in an error due to concentration changes that result. One way to partially compensate is to prepare a series of standard solutions using the sample matrix as the diluent. With either method, volumes of highly concentrated (or pure) analyte can be quite small (on the order of microliters) so that the dilution effect is negligible. A correction factor for the dilution can also be calculated. (Figure 5.6 shows the data from a GC experiment which was standard additions.)

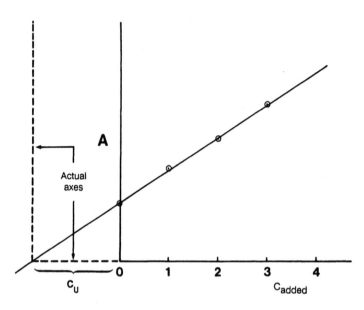

FIGURE 5.5   An example of a graph for a "standard additions" absorbance experiment.

5.4 EFFECT OF SAMPLE PRETREATMENT ON CALCULATIONS

The concentration obtained from the various graphs of the instrument readout (or ratio, etc.) vs concentration is rarely the final answer in an instrumental analysis. In most procedures, the sample has undergone some form of preanalysis treatment prior to the actual measurement. In some cases, the sample must be diluted prior to the measurement. In other cases, a chemical must be added prior to the measurement, possibly changing the analyte's concentration. In still other cases, the sample is a solid and must be dissolved prior to the measurement.

The instrument measurement is the measurement of the solution tested, and the concentration found is the concentration in that solution. What the concentration is in the original, untreated solution or sample must then be calculated based on what the pretreatment involved.

Frequently this is merely a "dilution factor." At other times, it is a calculation of the grams of the constituent from the molar concentration of the solution. Some examples of this follow. Refer to Chapter 3 for a discussion of the parts per million unit.

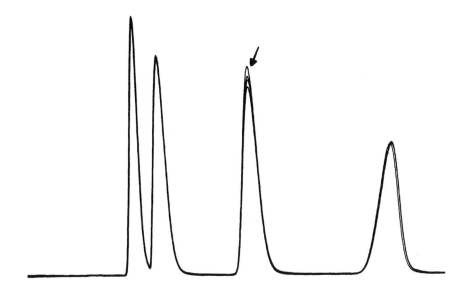

FIGURE 5.6   A series of chromatograms superimposed to illustrate standard additions in GC. The peak size of the analyte grows (arrow) while the others remain unchanged. (From Kenkel, J., *Analytical Chemistry for Technicians*, Lewis Publishers, Inc., Chelsea, MI, 1988. With permission.)

Example:

A water sample was tested for iron content, but was diluted prior to obtaining the instrument reading. This dilution involved taking 10 mL of the sample and diluting it to 100 mL. If the instrument reading gave a concentration of 0.891 ppm for this diluted sample, what is the concentration in the undiluted sample?

Solution:

$$\frac{0.891 \text{ mg}}{\text{L}} \times 0.100\text{L} = 0.0891 \text{ mg Fe in flask}$$

This is also the milligram of Fe in the 10 mL of undiluted sample. Thus,

$$\frac{0.0891 \text{ mg Fe}}{0.010 \text{ L water}} = 8.91 \text{ ppm Fe in original water}$$

Alternate Solution:
  The "Dilution Factor" is

$$\frac{0.100 \text{ L}}{0.010 \text{ L}} \quad \text{or} \quad \frac{100 \text{ mL}}{10 \text{ mL}} \quad \text{or } 10$$

  Therefore, $0.891 \times 10 = 8.91$ ppm Fe.

Example:
  A 1.000-g soil sample was analyzed for potassium content by extracting the potassium using 10.00 mL aqueous ammonium acetate solution. The soil was then rinsed, and the solution was diluted to exactly 50.00 mL. If the concentration in this 50 mL was found to be 5.27 ppm, what is the concentration of the potassium in the soil in ppm?
Solution:

$$\text{ppm K} = \frac{\text{mg K}}{\text{kg soil}} = \frac{5.27 \text{ mg/L} \times 0.05000 \text{ L}}{0.001000 \text{ kg}}$$

$$\frac{5.27 \text{ mg}}{\text{L}} \times 50 = 264 \text{ ppm in soil}$$

5.5 USE OF COMPUTERS

The role that computers play in the analytical laboratory is discussed both here and in Chapter 13. We will describe here four important functions that microprocessors, microcomputers, and computers in general perform in the analytical laboratory. These are (1) data acquisition, (2) data manipulation and storage, (3) graph plotting, and (4) experiment control.

5.5.1 Data Acquisition

It is very commonplace today for the analyst to use a computer for acquiring data. The term data acquisition refers to the fact that the data that is normally fed into a recorder (like so many of the instrumental techniques described in this text) can also simultaneously be fed into and acquired by a computer, perhaps even alongside the recorder and sharing the same output line from the instrument. Such data is thereby stored in short term in the computer's memory and, in the long term, on disk or tape. The actual experiment, however, requires the use of another piece of hardware most of the time because most computers will accept only "digital" signals, while the output signal from an instrument is most often of the continuous or "analog" variety. Thus one needs what is called an "analog to digital" ("A to D") converter, sometimes referred to as an "interface".* The continuously changing nature of an analog signal is evident on the chart recording of a chromatography peak or molecular absorption spectrum, for example. The A to D converter samples this data at predetermined time intervals and feeds these "digital" values into the computer where it is stored and/or displayed on the cathode-ray tube (CRT) screen. Individual discreet "digital" values are what the computer then sees.

The A to D converter, while described here as a separate piece of hardware, can be incorporated either into a circuit board designed to plug into a slot inside the microcomputer or within the instrument itself such that an output for computer data acquisition is provided. In either configuration, it appears externally that the computer is actually accepting an analog signal, which in a sense it is.

When in actual use, a program is often run on the computer which activates the A to D converter, establishes the sampling time interval and other parameters at the discretion of the operation, and begins the data acquisition at the touch of a key on the computer keyboard.

5.5.2 Data Manipulation and Storage

Any voltage signal generated by laboratory instruments and output to a recorder can be fed into a computer in the manner just described. This includes pH meters, spectrophotometers, chromatographs, polarographs, etc. Even automated analysis recorder signals can be fed into a computer. The real value of this, however, lies in the fact that the

---

*   The term "interface" is also used to describe the connection of any piece of hardware to a computer and not just an A to D converter.

data can be permanently stored on a computer disk or tape and recalled later for the purpose of manipulation, calculation, or whatever form of "reduction" is needed for a given type of data.

The value of being able to permanently store data on disk or tape is in the achievement of a "paperless" laboratory. Before the advent of the computer, huge cumbersome volumes of chart recordings had to be stored in the laboratory. Today, if the analyst wants to review data acquired even years earlier, he/she can call the data from disk or tape and view it in a matter of seconds.

5.5.3 Graph Plotting

Most instrumental analytical procedures require the plotting of a linear graph discussed earlier in this chapter for the purpose of establishing the exact relationship between the instrument reading and the concentration of the analyte. The least squares fit of the data points to the line is important in order to be as accurate as possible with the location of the line. Also, the calculation of the correlation coefficient is important. We have already indicated that a computer can be used to perform these functions. Indeed, the use of a computer to plot data, perform a least square fit, print out a copy of the fitted curve, and also calculate the correlation coefficient are very important computer functions in the modern lab. The improved accuracy of the determination can be demonstrated by manual plotting and curve fitting and comparing the answer to what a computer found. You will find that the speed of the determination also improves dramatically.

5.5.4 Experiment Control

Finally, entire experiments can be controlled with the use of microcomputers and microprocessors. Pressing a key on a computer keyboard can cause an experiment to begin, be altered, or to end. Indeed the more sophisticated modern instruments have the microprocessor for this built right into the instrument (essentially a microcomputer complete with disk drive and CRT and the instrument all in one unit) such that experiment control is quite easily accomplished.

Essentially any instrument can be told when to begin executing its function in this manner, even GC chromatographs in which the experiment begins when the operator pushes in the syringe plunger. An electrical contact closure can be incorporated into the syringe or injection port such that the syringe injection actually activates the computer for data acquisition.

CHAPTER 6

MOLECULAR SPECTROSCOPY

6.1 INTRODUCTION

More than half of all instrumental methods of analysis involve the measurement of either light absorption or light emission by a sample. For this reason, the analyst needs to have a basic understanding of the modern theory of light and parameters of light, namely wavelength, frequency, wave number, and energy, as well as the concept of how light interacts with matter and the theory behind the emission and absorption phenomena. This chapter introduces these concepts and continues with perhaps the oldest and some of the most popular of all instrumental analysis techniques, that of ultraviolet, visible, and infrared molecular absorption analysis. The final sections deal with NMR and mass spectrometry.

6.1.1 Nature and Parameters of Light

The modern theory of light says that light has a "dual nature." This means that light exhibits both the properties of waves (the wave theory) and the properties of particles (the particle theory). The wave theory says that light travels from its source via a series of repeating waves, much like waves of water move across the surface of a body of water. The particle theory says that light consists of a stream of particles called "photons" emanating from the source. For our purposes, the wave theory seems to have the most applicability.

The one important difference between waves of light and waves of water moving across a body of water is that the light waves are not mechanical waves like water waves. Light waves do not require matter to move or to exist. Rather than being mechanical disturbances, they are electromagnetic disturbances, and as such, they can travel through a vacuum, from the sun to the earth, for example. Since light waves are electromagnetic disturbances, light is often referred to as electromagnetic radiation. It has an electrical component and a magnetic component. Of particular importance in analytical chemistry are the wavelength, frequency, and energy of light as described by the wave theory.

Since light consists of a series of repeating waves, the physical distance from a point on one wave to the same point on the next wave is an important parameter. This distance is termed the wavelength. It is given the Greek symbol lower case lambda (λ) (see Figure 6.1). Wavelength can vary from distances as little as fractions of atomic diameters to several miles. This suggests the existence of a very broad spectrum of wavelengths. Indeed, the distinction between visible (vis) light, ultraviolet (UV) light, infrared (IR) light, etc. is the magnitude of the wavelength. Each of these, along with the others mentioned previously, encompass a particular "region" of the total electromagnetic spectrum. Thus, we have the vis region, the UV region, the IR region, etc. Figure 6.2 shows a representation of the entire electromagnetic spectrum, wavelength increasing left to right, and indicates the approximate wavelength borders of the various regions in nanometers. In the UV, vis, and IR regions, which are the regions that we will be emphasizing, the nanometer (nm) and the micrometer (or micron — μm) are the most commonly used units of wavelength.

Another parameter of light derived from the wave theory is frequency. Frequency is defined as the number of waves that pass a fixed point in 1 sec (cycles per second). It is given the Greek symbol nu (ν). Frequency would obviously depend on how fast the wave travels. However, all light travels at the same speed in a vacuum, 3.00×10^{10} cm/sec, the "speed

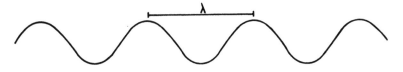

FIGURE 6.1   The definition of wavelength.

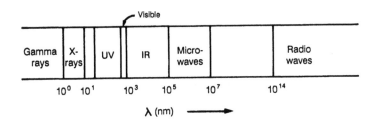

FIGURE 6.2   The electromagnetic spectrum.

of light." [*] Thus, a change in wavelength is the only change that would produce a different frequency. In other words, a particular wavelength corresponds to a particular frequency. Figure 6.3 should help clarify this concept. The most common unit of frequency is the hertz (hz) or sec^{-1}. Speed, wavelength, and frequency are mathematically related by the following equation.

$$\lambda = \frac{c}{v} \tag{6.1}$$

in which c is the speed of light, 3.00×10^{10} cm/sec.

A parameter closely related to frequency is the wave number. Wave number is defined as the reciprocal of the wavelength when the wavelength is expressed in centimeters and is given the symbol $\bar{v}$ (read "nu bar").

$$\bar{v} = \frac{1}{\lambda \text{ (cm)}} \tag{6.2}$$

---

[*] In a medium other than a vacuum, the speed of light is different from that in a vacuum. This creates a phenomenon known as refraction, a parameter known as refractive index, and a technique known as refractometry. These will be discussed in Chapter 10.

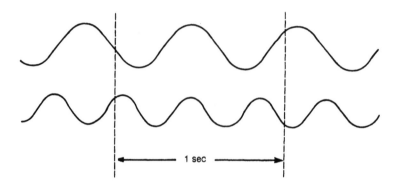

FIGURE 6.3   An illustration of how frequency changes with wavelength. A light with a shorter wavelength will have more waves (cycles) per second.

Wave number is an important parameter for the IR region, as we will see.

Last, but certainly not least, is energy. Different energies are associated with the different wavelengths and frequencies of light. Mathematically, we define energy as follows:

$$E = h\upsilon \tag{6.3}$$

and in terms of wavelength,

$$E = h\left(\frac{c}{\lambda}\right) \tag{6.4}$$

The symbol "h" is a proportionality constant known as "Planck's Constant" which has the value 6.62×10^{-27} erg sec/photon. The photon, as indicated previously, is the name given to a "particle" of light in the particle theory. The erg is a unit of energy in the metric system.

As we consider the theories of the interaction of UV, visible (vis), and infrared (IR) light with matter and the theories of molecular and atomic absorption in this chapter and the next, it is important to understand what the above four equations mean in terms of how one parameter relates to any of the others (see Table 6.1). In addition, we can indicate which regions of the electromagnetic spectrum are high energy regions, which are low energy regions, which are high frequency regions, which are low frequency regions, etc. (see Figure 6.4 for this). Concerning the three regions to be emphasized in this chapter, then, we can say the following:

Table 6.1   Statements of Interpretation of Equations 6.1, 6.2, 6.3, and 6.4

Equation	Interpretation
6.1	Frequency is inversely proportional to wavelength. As frequency increases, wavelength decreases and vice versa. Long wavelength corresponds to low frequency.
6.2	Wave number is inversely proportional to wavelength. As wavelength increases, wave number decreases and vice versa. Long wavelength corresponds to low wave number.
6.3	Energy is directly proportional to frequency. As frequency increases, energy increases and vice versa. High frequency corresponds to high energy.
6.4	Energy is inversely proportional to wavelength. As wavelength increases, energy decreases and vice versa. Long wavelength corresponds to low energy.

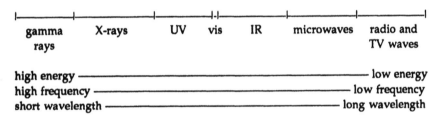

FIGURE 6.4   Comparisions of the various regions of the electromagnetic spectrum in terms of energy, frequency, and wavelength.

$$energy: \quad UV > vis > IR$$
$$wavelength: \quad UV < vis < IR$$
$$frequency: \quad UV > vis > IR$$

The visible region of the spectrum (vis) as shown in Figures 6.2 and 6.4 is a very "narrow" region compared to the others. Visible of course means that to which our human eye is sensitive. Thus it is apparent that we can see only a very small fraction of all the wavelengths of light that continuously criss-cross our atmosphere and universe. In addition, those individual wavelengths of visible light that enter our eye are characteristically interpreted, i.e., different wavelengths correspond to different colors. Thus we have red light, yellow light, green light, etc. Since the different colors correspond to different wavelengths, they also correspond to different energies and frequencies (see Figure 6.5).

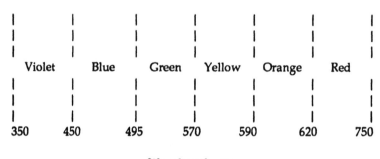

Wavelength, nm

FIGURE 6.5   The visible region of the electromagnetic spectrum and the approximate wavelength boundries of the colors.

6.1.2 Light Absorption and Emission

Evidence of light absorption abounds in our world. All colored objects constitute such evidence. Visible white light that comes from the sun, a light bulb, or any visible light source actually consists of all the visible wavelengths represented in Figure 6.5. Consider a rainbow, for example. White light coming to earth from the sun can, under the right conditions, be dispersed into the rainbow colors. The right conditions usually means a large concentration of water droplets in the atmosphere. This is evidence that white light consists of all the colors and that these colors are separable. Why does a red sheet of paper appear red? All the wavelengths of visible light from the visible light source in the room are absorbed except for the red wavelengths, and these are reflected to our eye. Why does a solution of potassium permanganate appear to be a deep purple color? The wavelengths of visible light which are incident on the solution from the light in the room are all absorbed except for those in the violet and red ends of the visible region. The result is an intense purple color.

How is it that light can be absorbed? What is happening in the molecular structure of a colored sample which results in the absorption of some wavelengths of light more than others? Elementary atomic and molecular theory gives us the answers. First, atoms and molecules, of which all matter is composed, can exist in a variety of energy levels depending on the state of their electrons. The electron clouds which are present outside the nucleus of atoms are layered in levels of various energy. Some of these levels, called "electronic" energy levels, have electrons in them and some do not. The reason for this is the varying number of electrons in the different atoms. No atom has enough electrons to fill all the levels. The electron configuration of lithium, for example, is $1s^22s^1$, while that of carbon is $1s^22s^22p^2$. The 2p level is present in a lithium atom, but, unlike carbon, does not have any electrons in it. Theoretically, all atoms have an infinite number of electron energy levels and so all atoms have many vacant levels above the level of the outermost electron. This is an important point because it means that atoms can "absorb" energy from an energy source so as to have its electrons promoted to higher vacant levels. Since a wavelength of light is associated with a certain amount of energy, this amount of energy in the form of light can be absorbed in order to promote an electron. Such a promotion is called an electronic energy transition. Thus, if light consisting of exactly the same energy as the difference between two electronic levels shines on a sample containing an atom which has these two levels, the light will "disappear" or be absorbed. The energy that once was light has been transferred to the atom. The original level, if it is the lowest possible level in which the electron can be found, is called the "ground state." All other levels are called "excited states" (see Figure 6.6).

Secondly, molecules exist in "vibrational" and "rotational" energy levels, as well as electronic levels. First consider vibrational levels. The bonds between the atoms in a molecule can behave as springs. When two balls connected by a spring are pulled apart and let go, the balls will move back and forth as the spring is stretched and contracted, constituting a "vibration" or, in the case of a molecular bond, a new vibrational energy level. Thus, a molecule can absorb energy resulting in the promotion to a higher vibrational energy level. Such a vibrational energy transition can mean a number of different vibrational modes for the bonds (see Figure 6.7). A molecule can therefore have a number of vibrational levels to which it can be promoted. As with the electronic energy level model, the energy needed to promote a molecule to a higher vibrational level can be provided by a wavelength of light, and this light can be absorbed, resulting in the promotion to this higher level. Similarly, a molecule can be made to rotate with the absorption of the appropriate energy, and thus there are also rotational energy levels.

Absorptions in the UV/vis/IR regions of the electromagnetic spectrum result in just these kinds of transitions — electronic, vibrational, and rotational. Electronic transitions are higher energy transitions (more energy required) than vibrational or rotational transitions. Electronic transitions are caused by UV and vis light; vibrational and rotational transitions are caused by IR light.

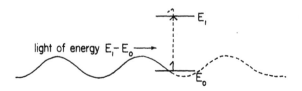

FIGURE 6.6   The absorption of light by an electron causing the promotion from an energy level E_0 (ground state) to an energy level E_1 (excited state). The energy that once was light now belongs to the electron.

FIGURE 6.7   Three vibrational modes for a 3-atom molecule. (From Kenkel, J., *Analytical Chemistry for Technicians*, Lewis Publishers, Inc., Chelsea, MI, 1988. With permission.)

   Vibrational levels are present in all electronic levels of molecules, and rotational levels are present in all vibrational levels. This means that when UV or vis light is absorbed, the transition can be from any vibrational or rotational level in any electronic level to any vibrational or rotational level in any higher electronic level. However, when IR light is absorbed, the transition can only be between two vibrational levels (short wavelength or "fundamental" IR) in the same electronic level or between two rotational levels within the same vibrational level (long wavelength or "far" IR). The initial electronic level is always the ground electronic state (see Figure 6.8). This places a greater restriction on the absorption possibilities for IR light and opens up a very large number of possibilities for UV/vis light. Only certain wavelengths of IR light have a chance of being absorbed for a particular molecule. However, many different wavelengths can be absorbed in certain parts of the UV and vis regions.

   These facts have an effect on the manner in which these techniques are used in the lab, as we will see. Similar statements can be made concerning the applicability of light absorption techniques to atoms as opposed to molecules. The absorption possibilities with atoms are greatly restricted due to the total absence of vibrational and rotational levels. However, "atomic" absorption techniques are no less important than molecular techniques, as should be evidenced by the discussions in Chapter 7.

   Finally, emission of light by a molecular sample can also occur. The term for such an emission is "fluorescence." Fluorescence occurs when

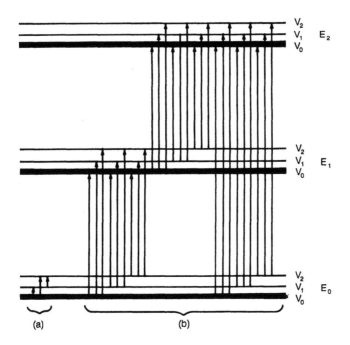

V_2
V_1 E_2
V_0

V_2
V_1 E_1
V_0

V_2
V_1 E_0
V_0

(a)   (b)

FIGURE 6.8   (a) The vibrational transitions possible in the first three vibrational levels of a molecule. (b) The electronic transitions possible in a molecule's first three electronic levels in which there are three vibrational levels each.

an electron that has been promoted to an excited state loses the energy it has gained and drops back to a lower state, although not all the way back to the original state. The energy that is lost is on the order of light energy, but light of a different wavelength than was originally absorbed, since it does not drop all the way back. The sample thus appears to "glow" due to the different wavelength. An analysis technique utilizing this phenomenon, called fluorometry, will be described later in this chapter.

6.2 NAMES OF TECHNIQUES AND INSTRUMENTS

The absorption of light at particular wavelengths is obviously an important measurement. It is therefore important for all instruments designed to measure such absorption to have, as part of the absorption measurement process, the capability of resolving an entire spectral region (polychromatic light) into these particular wavelengths ("monochromatic" light). Two general terms which describe all techniques involving light, including those requiring monochromatic light as just described, are "spectroscopy" and "spectrometry." The general term for the instrument used is "spectrometer." Any spectrometer that resolves polychromatic light into monochromatic light and utilizes a "photomultiplier" tube (described later) as a detector of light is properly called a "spectrophotometer" and the corresponding technique "spectrophotometry." This latter term correctly applies to UV and vis instruments, but technically not to IR instruments, since in that case, as we will see, the detector is not a photomultiplier tube. Nonetheless, IR spectrometers are often referred to as spectrophotometers. It is most correct to refer to an IR instrument as a spectrometer and to the technique as IR spectrometry. In addition, an instrument constructed only to be used in the vis region is often called a colorimeter, since the monochromatic light and the solutions measured would display a color. The technique which utilizes a colorimeter is called colorimetry. This term is also often used to describe a technique in which the color of an unknown is visually matched (rather than with an instrument) to a set of standards in order to determine concentration.

6.3 UV/vis SPECTROPHOTOMETRY

6.3.1 General Description

We will first study the analytical technique involving the absorption of UV and vis light. This technique, like the IR technique to be discussed in the next section, involves the use of a laboratory instrument that measures the degree of light absorption by a sample. In addition to the detector (Section 6.2), there are other features that distinguish the UV/vis instruments from the IR instruments, as we will see. The design of the UV/vis instrument is such that a monochromatic wavelength of light from a light source (inside the instrument) is allowed to strike a sample solution. The amount of light absorbed by this solution is electronically measured by a photomultiplier tube and displayed on a readout device. The analyst is able to quantitate a constituent in the sample by relating this degree of absorbance displayed by the instrument to the constituent's concentration. In addition to this quantitative analysis, a qualitative analysis can also be performed by observing the pattern of absorption that a sample exhibits over a range of wavelengths, the so-called "molecular absorption spectrum." No two such patterns from any two chemical species are exactly alike. We therefore have what can be called a molecular "fingerprint," and this is what makes identification, or qualitative analysis, possible. The modern instrument is capable of both the qualitative and quantitative measurements. The specific schemes by which qualitative and quantitative analysis is accomplished will be described, but let us begin by detailing the inner workings of the instrument.

6.3.2 Instrument Design

There are two types of UV/vis instruments in common use: the single-beam instrument and the double-beam instrument. These instruments have many common features and utilize the same components. First, the single-beam instrument and its components will be described, and this will be followed by a description of the double-beam instrument. A diagram of the basic single-beam spectrophotometer is shown in Figure 6.9. The light source (polychromatic) provides the light to be directed at the sample. The wavelength selector, or monochromator, isolates the wavelength to be used. The sample holder/compartment is a light-tight "box" in which the sample solution is held, and the detector/readout components are the electronic modules which measure and display the degree of absorption. Let us describe each of these components in more detail.

6.3.2a The Light Source

The light source for the vis region is different from the light source for the UV region. Instruments which have the capability of measuring the absorption in both regions must have two independently selectable light sources. For the vis region, a light bulb with a tungsten filament is used. Such a source is very bright and emits light over the entire vis region and into the near IR region. The intensity of the light varies dramatically across this wavelength range (Figure 6.10). This creates a bit of a problem for the analyst, and we will discuss this later.

For the UV region, the light source is usually the deuterium discharge

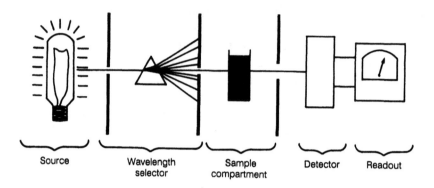

| Source | Wavelength selector | Sample compartment | Detector | Readout |

FIGURE 6.9   The essential components of a spectrophotometer.

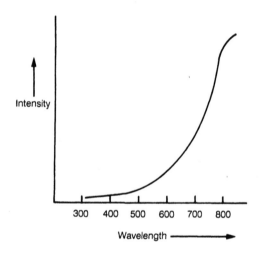

FIGURE 6.10   A graph showing approximately how the intensity of light from a tungsten filament source varies according to wavelength.

lamp. Its wavelength output ranges from 185 to about 375 nm, satisfactory for most UV analyses. Here again the intensity varies with wavelength.

6.3.2b The Monochromator

The monochromator, or wavelength selector, consists of three main parts: an entrance slit, a dispersing element, and an exit slit. In addition to these, there is often a network of mirrors situated for the purpose of aligning or collimating a beam of light before and after it contacts the dispersing element. A slit is a small circular or rectangular opening cut into an otherwise opaque plate, such as a painted metal plate. The size of the opening is often variable — a "variable slit width". The entrance slit is where light enters the monochromator from the source (see Figure 6.11). Its purpose is to create a unidirectional beam of light of appropriate intensity from the multidirectional light emanating from the source. Its slit width is usually variable so that the intensity of the beam can be varied; the wider the opening the more intense the beam.

After passing through the entrance slit, the beam strikes a dispersing element. The dispersing element disperses the light into its component wavelengths. For vis light, for example, this would mean that a beam of white light is dispersed into a spray of rainbow colors, the violet/blue wavelengths on one end to the red wavelengths on the other, with the green and yellow in between. The monochromatic light is then selected by the exit slit. As the dispersing element is rotated, the spray of colors moves across the exit slit such that a particular wavelength emerges at each position of rotation. The exit slit width can be variable too, but making it wider would result in a wider wavelength band (the "band pass") passing through, which is usually undesirable. The light emerging from this exit slit is therefore monochromatic and is passed on to the sample compartment to strike the sample. The concept is the same for UV light (see Figure 6.11). The rotation of the dispersing element is accomplished by either manually turning a knob on the face of the instrument or internally by programmed scanning controls. The position of the knob is coordinated with the wavelength emerging form the exit slit such that this wavelength is read from a scale of wavelengths on the face of the instrument or on a readout meter. For manual control, the operator thus simply dials in the desired wavelength.

The dispersing element is either a diffraction grating or a prism. A prism is a three-dimensional triangularly shaped glass or quartz block as indicated in Figure 6.9. When the light beam strikes one of the three faces of the prism, the light emerging through another face is dispersed.

A diffraction grating is used more often than a prism. A diffraction grating is like a highly polished mirror that has a large number of precisely parallel lines or grooves scribed onto its surface. Light striking this surface is reflected, diffracted, and dispersed into the component wavelengths as indicated in Figure 6.11. See Figure 6.12 for more explicit diagrams of a prism and a diffraction grating.

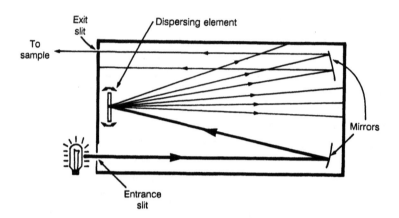

FIGURE 6.11   The monochromator component (see text for full description).

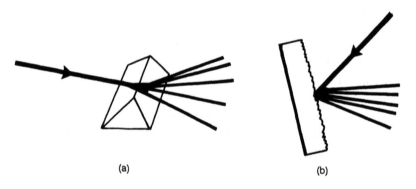

FIGURE 6.12   (a) A prism and (b) a diffraction grating.

6.3.2c Sample Compartment

Next, the light from the monochromator passes through the sample in the sample compartment. For the UV/vis instrument, this is a light-tight box in which the container holding the sample solution is placed. The container is called a "cuvette." The material making up the cuvette walls is of interest.

For optimum performance, this material must be transparent to all wavelengths of light that may pass through it. For vis light, this means that the material must be completely clear and colorless. Inexpensive materials, such as plastic and ordinary glass, are perfectly suitable in that case. However, such materials may not be transparent to light in the UV and IR regions. For the UV region, the more expensive quartz cuvettes must be used, while in the IR region, the walls of the cuvette are often made of inorganic salt crystals (see Section 6.4).

6.3.2d Detector/Readout

The most common detector for UV/vis spectrophotometry is the photomultiplier tube, which is comparable to a solar cell. When light of a particular intensity strikes it, a current, or electrical signal, of proportional magnitude is generated, amplified, and sent on to the readout. The process is not energy efficient like a solar cell needs to be, however, since it requires more electricity to run it than it generates.

The photomultiplier tube consists of a "photocathode," an anode, and a series of "dynodes" for multiplying the signal, hence its name. The dynodes are situated between the photocathode and the anode. A high voltage is applied between the photocathode (an electrode which emits electrons when light strikes it) and the anode. When the light beam from the sample compartment strikes the photocathode, electrons are emitted and accelerated, because of the high voltage, to the first dynode where more electrons are emitted. These electrons pass on to the second dynode, where even more electrons are emitted, etc. When the electrons finally reach the anode, the signal has been sufficiently multiplied as to be treated as any ordinary electrical signal able to be amplified by a conventional amplifier. This amplified signal is then sent on to the readout in one form or another. Figure 6.13 illustrates this process.

Most instruments offer a choice as to what is displayed on the readout, since there are several parameters that can have analytical importance or convenience. These parameters, notably "transmittance," "percent transmittance," "absorbance," and analytical concentration, will be defined and discussed. The display of each of these requires associated mathematical operations so as to obtain it from the intensity of the light to which the detector's output signal is proportional. Thus, electronic circuits which compute a ratio of intensities (transmittance), the percent transmittance, the negative logarithm of this ratio (absorbance), etc. are also often included as part of the total circuitry of these instrument components. In addition, the signal may be digitized, using an analog to digital converter (Chapter 5) for display on a digital readout or for output to a computer.

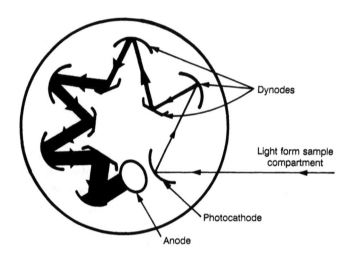

Dynodes

Light form sample
compartment

Photocathode

Anode

FIGURE 6.13   The process occurring in a photomultiplier tube.

6.3.2e Calibration

The calibration of the single-beam instrument that has just been described is important to consider. Such calibration involves adjustment of
the "dark current" control when all the light is physically blocked from
striking the detector and adjustment of the entrance slit opening to the
optimum value when no analyte species is present in the path of the light.
In order to understand these two steps, we need to define some parameters involved.

The intensity of light striking the detector when a "blank" solution (no
analyte species) held in the cuvette is present is given the symbol "I_0."
The blank is a solution that contains all chemical species that will be
present in the standards and samples to be measured (at equal
concentration levels) except for the analyte species. Such a solution should
not display any absorption and thus I_0 represents the maximum intensity
that can strike the detector at any time. When the blank is replaced with
a solution of the analyte, a less intense light beam will be detected. The
intensity of the light for this solution is given the symbol "I" (see Figure
6.14). The fraction of light transmitted is thus I/I_0. This fraction is defined
as the "transmittance," "T."

$$T = \frac{I}{I_0}$$

(6.5)

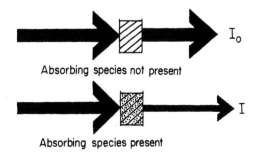

Absorbing species not present

Absorbing species present

FIGURE 6.14   Illustration of the definitions of I and I_0. (From Kenkel, J., *Analytical Chemistry for Technicians*, Lewis Publishers, Inc., Chelsea, MI, 1988. With permission.)

The "percent transmittance" is similarly defined.

$$\%T = T \times 100 \tag{6.6}$$

Transmittance and percent transmittance (%T) are two parameters that are able to be displayed on the readout. Calibration thus involves adjustment to 0% T using the dark current control when the light is physically blocked from the detector and adjustment to 100% T using the entrance slit opening control when the blank solution is in the path of the light. Opening or closing the entrance slit increases or decreases the light intensity eventually striking the detector and thus is useful for such calibration. Following calibration, the samples and standards, in the same or identical cuvettes (in order to measure only the effect of the solution), are consecutively placed in the path of the light and the readout recorded.

6.3.2f Double-Beam Instruments

The discussion in this section thus far has been concerned with instruments in which a single beam of monochromatic light passes through the sample compartment. While such instruments are in very common use (they are comparatively inexpensive), an instrument design which utilizes two beams of the light passing through the sample compartment is also quite common and offers some important advantages.

In any spectrophotometer, if the intensity of the light passing through the sample changes following calibration and before a sample or standard is read, an error will result. Such a change can occur due to fluctuations in the power supply to the source and/or detector (the line voltage), due to an unstable source, or when the operator selects a different wavelength via the monochromator control. The intensity of the light changes dramatically as the wavelength through the monochromator is changed (refer back to Figure 6.10). If a given experiment involves a change in the wavelength, such as when determining the pattern of absorption over a range of wavelengths (the "molecular absorption spectrum"), or if one suspects that the line voltage fluctuates, recalibration with the blank is necessary periodically or each time the wavelength is changed. This can result in an extraordinarily long and tedious procedure and/or a loss of accuracy, since the procedure of recalibration with the blank can take up to 5 to 10 sec and a power fluctuation can occur in this length of time. The double-beam design allows the blank to be checked and calibration to take place only a split second before the sample is read. Not only does this take much less time, eliminating the need for continuous manual monitoring of the blank, but it also increases accuracy, since the time between blank calibration and sample measurement is dramatically decreased. It also allows rapid wavelength scanning in order to conveniently obtain the molecular absorption spectrum.

A schematic diagram of a typical double-beam design for UV/vis spectrophotometry is shown in Figure 6.15. The light coming from the monochromator is directed along either one of two paths with the use of a "chopper." The chopper in this case is a rotating circular half-mirror used for splitting a light beam into two beams. At one moment, the light passes through the sample, while at the next moment it passes through the blank. Both beams are joined again with a beam combiner, such as another rotating half-mirror, prior to entering the detector. The detector sees alternating light intensities, I and I_0, and thus immediately and automatically compensates for fluctuations and wavelength changes, usually by automatically widening or narrowing the entrance slit to the monochromator. If the beam becomes less intense, the slit is opened; if the beam becomes more intense, the slit is narrowed. Thus the signal relayed to the readout is free of effects of intensity fluctuations that cause errors.

Sample compartments in such instruments have two cuvette holders, one for a cuvette containing the sample or standard and one for a cuvette containing the blank. The two beams of light pass through the sample compartment, one through the blank (the reference beam) and one through the sample or standard (the sample beam). The two cuvettes must be matched in terms of pathlength and reflective and refractive properties.

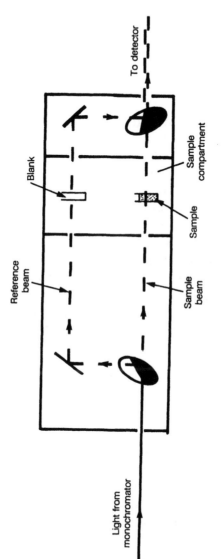

FIGURE 6.15   A schematic diagram of a typical double-beam design.

Scanning double-beam UV/vis spectrophotometers have become very commonplace in analytical laboratories. Modern instruments can include substantial microprocessor control, including control of scan time, output functions, and data storage, as well as internal storage of Beer's Law data such that sample concentrations can be displayed on the readout. These instruments also have output terminals capable of transferring concentration and absorption/transmittance and wavelength data to an external recorder or computer.

6.3.3 Qualitative and Quantitative Analysis

The transmittance and percent transmittance of a sample have been defined (Equations 6.5 and 6.6). The unfortunate aspect of transmittance is that it is not linear with concentration. Low concentrations give a high transmittance, and high concentrations give a low transmittance. However, the relationship is not linear but logarithmic (see Figure 6.16a). Without a linear relationship, the concentration of an unknown sample cannot be determined by the usual procedures outlined in Chapter 5. However, since the relationship is logarithmic, the logarithm of the transmittance can be expected to be linear. Thus, we define the parameter of "absorbance" as being the negative logarithm of the transmittance and give it the symbol "A":

$$A = -\log T \qquad (6.7)$$

Absorbance is a parameter then that increases linearly with concentration (see Figure 6.16b). Absorbance is thus the parameter that is important for quantitative analysis. If transmittance is measured by an instrument, it must be converted to absorbance via Equation 6.7 (or by semilog graph paper) before plotting vs concentration and obtaining the unknown concentration according to procedures outlined in Chapter 5. However, as indicated previously, most instruments have the electronic circuitry for calculating absorbance built into the detector/readout system, and thus are capable of displaying absorbance on the readout, including on a recorder, such as for a display of the molecular absorption spectrum. Such a display is useful for a qualitative analysis, since, as indicated previously, it is a molecular fingerprint of the system studied. Let us study quantitative and qualitative analysis in UV/vis spectrophotometry beginning with this latter concept.

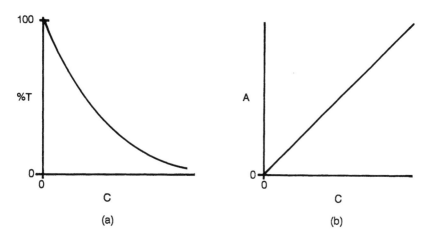

FIGURE 6.16   (a) A plot of percent transmittance vs concentration. (b) A plot of absorbance vs concentration.

6.3.3a Qualitative Analysis

The molecular absorption spectrum, the pattern of absorption over a range of wavelengths, and the molecular "fingerprint" of a particular chemical species, has been referred to at several junctures in this chapter. We are now ready to expand on this concept and look at some examples. The absorption pattern can be displayed as either a plot of absorbance vs wavelength or transmittance vs wavelength. Since absorbance and transmittance are complimentary, the transmission spectrum appears as an inversion of the absorption spectrum. Some examples of absorption spectra are given in Figure 6.17 and some examples of transmission spectra in Figure 6.18. All represent molecular fingerprints — no two chemical species will display identical absorption and transmission spectra. Qualitative analysis can simply involve a matching-up of spectra, known with unknown.

Absorption in the UV region warrants additional comment. Certain characteristics of the structure of organic compounds will display certain unique characteristics in UV absorption spectra. In the first place, organic structures with nothing but single bonds, such as alkanes and ordinary alcohols, do not absorb at all in the UV region. In fact, they are often used as solvents for compounds that do absorb. However, structures with double bonds, triple bonds, or benzene rings have very strong absorption patterns. All of these types of structures contain π electrons, and π electrons are especially susceptible to UV absorption. Any such group present

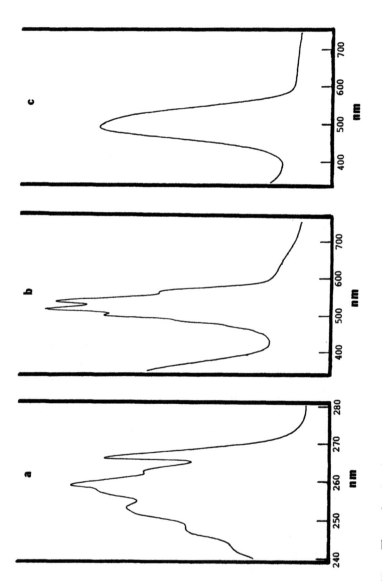

FIGURE 6.17   The molecular absorption spectra of (a) toluene in cyclohexane, (b) potassium permanganate in water, and (c) methyl red in water.

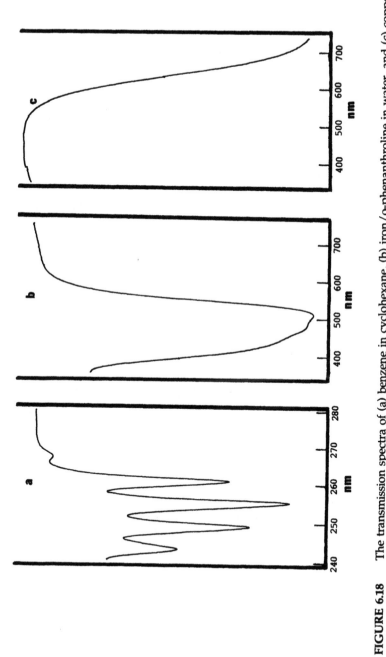

FIGURE 6.18   The transmission spectra of (a) benzene in cyclohexane, (b) iron/o-phenanthroline in water, and (c) copper sulfate in water.

in a structure, causing the appearance of absorption bands in the UV (or vis) region, is called a "chromophore." A –C=C– bond, a –C=O bond, a –N=O bond, and a benzene ring are thus all classified as chromophores.

If two or more chromophores appear in the same molecule, the location of one relative to the other dictates a particular pattern. If they are separated by more than one carbon, the absorption pattern represents a simple summation of the two individual patterns. If the two chromophores are on adjacent carbons, an apparent "shifting" (called a "bathochromic" shift) of the absorption maximum to a longer wavelength is usually observed, and the absorbance is increased (a "hyperchromic" effect). If the opposite is observed (a shift to a shorter wavelength), it is called a "hypsochromic" effect. When the two chromophores are attached to the same atom, the two observations just noted also occur, but to a lesser extent. Additional shifting of the absorption maximum can occur when an ordinarily nonabsorbing group is attached near a chromophore. Such a group is called an "auxochrome." An example would be an –OH group attached near a double bond. Additionally, effects of solvent and pH are important to observe. Thus, in addition to fingerprinting (spectra matching), observations of shifting of absorption maxima and absorbance intensity can be very important in a qualitative analysis.

6.3.3b Quantitative Analysis

The exact relationship between absorbance and concentration is a famous one. It is known as the Beer-Lambert Law, or simply as Beer's Law. A statement of Beer's Law is

$$A = abc \qquad (6.8)$$

in which A is absorbance, a is "absorptivity" or "extinction coefficient," b the "pathlength," and c the concentration. Absorbance was defined previously in this section. Absorptivity is the inherent ability of a chemical species to absorb light and is constant at a given wavelength. Pathlength is the distance the light travels through the measured solution. It is the inside diameter of the cuvette. Pathlength is measured in units of length, usually centimeters or millimeters. Concentration can be expressed in a variety of units, usually, however, in molarity, parts per million, or grams per 100 mL. The units of absorptivity depend on the units of these other parameters, since absorbance is a dimensionless quantity. When the concentration is in molarity and the pathlength is in centimeters, the units of absorptivity must be liters/(mole centimeter). Under these specific conditions, the absorptivity is called the "molar absorptivity," or the "molar extinction coefficient," and is given a special symbol, the Greek letter epsilon, (ε). Beer's Law is therefore sometimes given as

$$A = \varepsilon bc \qquad (6.9)$$

Defining the molar absorptivity parameter presents analytical chemists with a standardized method of comparing one spectrophotometric method with another. The larger the molar absorptivity, the more sensitive the method. (It is not unusual for molar absorptivity values to be as large as 10,000 L/(mol cm) and higher.) In addition, it was stated above that the absorptivity is constant "at a given wavelength," implying that it changes with wavelength. The greatest analytical sensitivity occurs at the wavelength at which the absorptivity is a maximum. This is the same wavelength that displays the maximum absorbance in the molecular absorption spectrum for that species. In fact, the molecular absorption spectrum is sometimes shown as a plot of absorptivity vs wavelength rather than absorbance vs wavelength. For a given absorbing species, such a plot would display the same characteristic shape and the same wavelength of maximum of absorbance.

Beer's Law constitutes the spectrophotometry application of the discussion in Chapter 5 which states that in most instrumental quantitative analyses, an instrument readout is proportional to concentration. In this case, absorbance is the readout. Thus, most quantitative analyses by Beer's Law involve preparing a series of standard solutions, measuring the absorbance of each in identical cuvettes, and plotting the measured absorbance vs concentration, creating the so-called "standard curve." The absorbance of an unknown solution is then measured and its concentration determined from the graph. Such a graph is often called a Beer's Law Plot (see Figure 6.19).

Of course, an unknown's concentration can be determined by comparing its absorbance with just one standard,

$$\frac{c_u}{c_s} = \frac{A_u}{A_s} \qquad (6.10)$$

in which c_u is the concentration of the unknown, c_s is the concentration

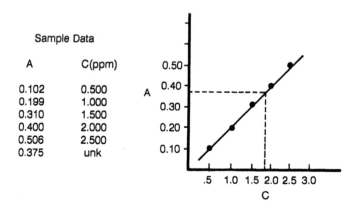

Sample Data

A	C(ppm)
0.102	0.500
0.199	1.000
0.310	1.500
0.400	2.000
0.506	2.500
0.375	unk

FIGURE 6.19   Some sample data and a Beer's Law Plot of the data showing the determination of the unknown concentration.

of the standard, A_u is the absorbance of the unknown, and A_s is the absorbance of the standard. It may also be determined by direct calculation, if the absorptivity value and pathlength are precisely known.

$$c = \frac{A}{ab} \tag{6.11}$$

See Section 5.2 for a more thorough discussion of each of these methods, including limitations, and also the calculations that are often involved after the unknown solution concentration is determined.

6.3.3c Interferences and Deviations

Interferences are quite common in qualitative and quantitative analysis by UV/vis spectrophotometry. An interference is a contaminating substance that gives an absorbance signal at the same wavelength or wavelength range selected for the analyte. For qualitative analysis, this would show up as an incorrect absorption spectrum, thus possibly leading to erroneous conclusions if the contaminant was not known to be present. For quantitative analysis, this would result in a higher absorbance than one would measure otherwise. Absorbances are additive. This means that the total absorbance measured at a particular wavelength is the sum of absorbances of all absorbing species present. Thus, if an interference is present, the correct absorbance can be determined by subtracting the absorbance of the interference at the wavelength used, if it is known. The modern solution to these problems is to utilize separation procedures, such as extraction or liquid chromatography, to separate the interfering substance from the analyte prior to the spectrophotometric measurement. These techniques will be discussed in Chapters 8 and 10.

Deviations from Beer's Law are in evidence when the Beer's Law Plot is not linear. This is probably most often observed at the higher concentrations of the analyte (see Figure 6.20). Such deviations can be either chemical or instrumental. Instrumental deviations occur because it is not possible for an instrument to be accurate at extremely high or extremely low transmittance values — values that are approaching either 0% T or 100% T. The normal working range is between 15 and 80%, corresponding to absorbance values between 0.10 and 0.82. It is recommended that standards be prepared so as to measure in this range and so that unknown samples be diluted if necessary. Chemical interferences occur when a high or low concentration of the analyte causes chemical equilibrium shifts in the solution which directly or indirectly affect its absorbance. It may be necessary in these instances to work in a narrower concentration range than expected. This means that unknown samples may also need to be further diluted as in the instrumental deviation case.

6.4 IR SPECTROMETRY

6.4.1 Introduction

As discussed in Section 6.1, the absorption of IR light causes vibrational energy transitions in molecules. The usefulness of the technique lies especially in the fact that only very specific wavelengths (energies) of IR light are able to be absorbed when a single particular kind of molecule is in the path of the light. As with UV/vis absorption, the absorbance vs wavelength plot is a molecular fingerprint of the molecule. The difference, however, lies in the fact that in the IR region, the absorption bands are extremely sharp, and each such band is associated with a particular covalent bond present in the molecule. We referred earlier to the ball and

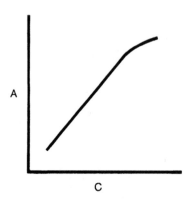

FIGURE 6.20   A Beer's Law Plot that shows a deviation from linearity at higher concentrations.

spring model of a molecule and how the initiation of a vibration of the spring constitutes an energy transition caused by an IR wavelength. The point here is that the wavelength, at which the sharp absorption band is observed, depends on what atoms the "spring" connects and whether the bond is a single bond, a double bond, etc. Thus, a –C–H– bond absorbs a particular wavelength, a –C–O– bond a different wavelength, a –C=O yet a different wavelength, etc. Additionally, various wavelengths can be absorbed depending on what mode of vibration is involved. These can be stretching vibrations, rocking and bending motions, etc. (refer back to Figure 6.7). The technique is especially useful therefore for qualitative analysis. Not only is the IR absorption spectrum a molecular fingerprint, but the presence of particular bonds in the structure will be manifested in corresponding sharp absorption bands at particular wavelengths, a fact that has considerable usefulness in the narrowing down of the possibilities in a qualitative analysis scheme. This topic will be expanded upon later in this section.

The nature of the sample holder is important to consider. In Section 6.3, we referred to the need for the sample holder in IR spectrometry to be composed of inorganic salt crystals. Since IR wavelengths are of such energy so as to cause covalent bonds to vibrate, any covalent compound would not be suitable as sample holder material because such material will absorb its characteristic wavelengths of IR light. This, in turn, may obviously cause erroneous conclusions in qualitative analysis schemes. Glass or plastic materials therefore cannot be used. Material with ionic bonds, however, is not a problem because ionic bonds do not absorb IR wavelengths — they do not undergo vibrational energy transitions. Thus, inorganic compounds, such as NaCl and KBr, are commonly used as matrix material and sample "windows" in IR spectrometry.

6.4.2 Liquid Sampling

For pure liquids ("neat" liquids) and liquid solutions, sandwiching a thin layer of liquid between two large NaCl or KBr crystals (windows) is the classic procedure for mounting the sample in the path of the light. Typical dimensions for such windows is about 2 cm wide × 3 cm long × 0.5 cm thick. Positioning or holding the crystals in place is done using either a "sealed cell," a "demountable cell," or a combination "sealed demountable" cell. "Sealed" cells are permanent fixtures for the windows and cannot be disassembled. They have a fixed pathlength and are very useful for quantitative analysis, since the pathlength is reproduced. "Demountable" cells can be disassembled so as to change the pathlength. The sample is placed on the window in a space created by a "spacer" while it is disassembled. The thickness of the spacer establishes the pathlength. It is then reassembled with the thin layer in place between the windows. "Sealed demountable" are demountable cells which include inlet and outlet ports for introducing and eliminating the liquid samples. All demountable cells are designed to allow easy disassembly for changing the spacer between the crystals (the pathlength) or for eliminating the spacer altogether for viscous samples, for example. Figure 6.21 shows a drawing of a typical sealed demountable cell. The top neoprene gasket and window have holes drilled in them to coincide with the inlet and outlet ports to facilitate filling the space, created by the spacer, with the liquid sample. The path of the liquid sample is shown as a dashed line in the figure.

Filling the cells with sample and eliminating the sample when finished can be troublesome. The sample inlet and outlet ports are tapered to receive a syringe with a Luer (tapered, ground-glass) tip. The usual procedure for filling is to raise the outlet end by resting it on a pencil or similar object (to eliminate air bubbles) and then to use a pressure/ vacuum system with the use of two syringes, one in the outlet port and one in the inlet port, as shown in Figure 6.22. While pushing on the plunger of the syringe containing the liquid sample in the inlet port and pulling up on the plunger of the empty syringe in the outlet port, the cell can be filled without excessive pressure on the inlet side. This reduces

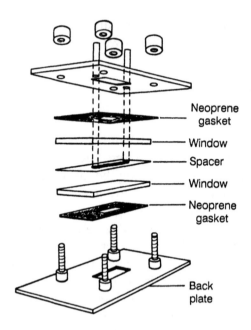

FIGURE 6.21 The "sealed demountable" cell assembly. (Adapted from Chia, L. and Ricketts, S., *Basic Techniques and Experiments in Infrared and FT-IR Spectroscopy*, The Perkin-Elmer Corporation, Norwalk, CT, 1988.)

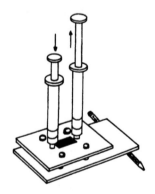

FIGURE 6.22 The recommended method of filling a sealed or demountable cell.

the possibility of damaging the cell due to the excessive pressure that may be needed, especially when working with unusually viscous samples and short pathlengths. Tapered Teflon plugs are used to stopper the ports immediately after filling. The cell may be emptied and readied for the next sample by using two empty syringes and the same push-pull method. When refilling, an excess of liquid sample may be used to rinse the cell and eliminate the residue from the previous one. Alternatively, the cell may be rinsed with a dry volatile solvent and the solvent evaporated before introducing the next sample.

The analyst must be careful to protect the salt crystals from water during use and storage. Sodium chloride and potassium bromide are, of course, highly water soluble, and the crystals may be severely damaged with even the slightest contact with water. All samples introduced into the cell must be dry. This is important for another reason, of course. Water contamination will show up on the measured spectrum and cause erroneous conclusions. If the windows are damaged with traces of water, they will become "fogged" and will appear to become nontransparent. The windows may be repolished if this happens. Depending on the extent of the damage, various degrees of abrasive materials may be used, but the final polishing step must utilize a polishing pad and a very fine abrasive. Polishing kits are available for this. Figure 6.23 shows the correct method for polishing. Finger cots should be used to protect the windows from finger moisture.

Liquids can be sampled as either the neat liquid (pure) or mixed with a solvent (solution). The neat liquid is more desirable since the spectrum will show absorption due to the liquid only. However, when a solution is run, both the analyte and the solvent will produce absorption bands, and they must be differentiated. Some solvents have rather simple IR spectra and are thus desirable for solvents. Examples are carbon tetrachloride (only –C–Cl bonds) and methylene chloride (CH_2Cl_2). Their IR spectra are shown in Figure 6.24. Alternatively, a double-beam instrument can be used to cancel out the solvent absorption. These instruments will be discussed later in this section.

6.4.3 Solid Sampling

IR spectrometry is one of the few analytical techniques that routinely analyzes solid undissolved samples. The techniques to be described here include the KBr pellet, the Nujol (mineral oil) mull, and the diffuse reflectance method.

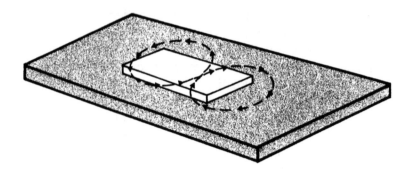

FIGURE 6.23   The correct method for polishing crystal materials for infrared sample cells. Note the figure eight motion.

6.4.3a KBr Pellet

The KBr pellet technique is based on the fact that *dry*, finely powdered potassium bromide has the property of being able to be squeezed under very high pressure into transparent discs — transparent to both IR light and vis light. It is important for the KBr to be dry both in order to obtain a good pellet and to eliminate absorption bands due to water in the spectrum. If a small amount of the dry solid analyte (0.1 to 2.0%) is added to the KBr prior to pressing, then a disc (pellet) can be formed from which a spectrum of the solid can be obtained. Such a disc is simply placed in the path of the light in the instrument and the spectrum measured.

Two methods of pressing the KBr pellet will be described here. First, a pellet die consisting of a threaded body and two bolts with polished faces may be used. One bolt is turned completely into the body of the die. A small amount of the powdered sample, enough to cover the face of the bolt inside, is added, and the other bolt is turned down onto the sample, squeezing it into the pellet (see Figure 6.25). The two bolts are then carefully removed. The body of the die is placed in the instrument so that the light beam passes directly through the center of the die.

The other method utilizes a hydraulic press and a cup die. The cup die consists of a base, with a protruding center, and a hollow cylinder that fits snugly over the protrusion. With the cylinder in place, the powdered sample is added to the cylinder to cover the face of the protrusion, and a second metal piece with a protrusion is placed on the top of the assembly. The entire assembly is placed into a laboratory press (see Figure 6.26).

It is important in either case for the KBr and sample to be dry, finely powdered, and well mixed. An agate mortar and pestle is recommended for the grinding and mixing of the KBr and sample.

6.4.3b Nujol Mull

The Nujol (mineral oil) mull is also often used for solids. In this method, a small amount of the finely divided solid analyte (1–2 μm particles) is mixed together with an amount of mineral oil to form a mixture with a toothpaste-like consistency. This mixture is then placed (lightly squeezed) between two NaCl or KBr windows which are similar to those used in the demountable cell discussed previously for liquids. If the particles of solid are not already the required size when received, they must be finely ground with an agate mortar and pestle and can be ground directly with mineral oil to create the mull to be spread on the window. Otherwise, a small amount (about 10 mg) of the solid is placed on one window along with one small drop of mineral oil. A gentle rubbing of the two windows together with a circular or back and forth motion creates the mull and distributes it evenly between the windows. The windows are placed in the demountable cell fixture and placed in the path of the light.

A problem with this method is the fact that mineral oil is a covalent compound and its characteristic absorption spectrum will be found superimposed in the spectrum of the solid analyte, as with the solvents used for liquid solutions discussed previously. However, the spectrum is a simple one (Figure 6.27) and often does not cause a significant problem.

6.4.3c Diffuse Reflectance

With the advent of the Fourier Transform Infrared Spectrometer (FTIR) instrumentation (see below), a technique called diffuse reflectance is becoming popular for solids. In this technique, the powdered KBr/analyte mixture (about 5% analyte) is placed in a small sample cup, and the light beam shines directly on this powdered sample. The diffuse reflectance, or the light returning from the surface of the sample that is scattered (and not simply reflected), is measured by the FTIR system and the spectrum

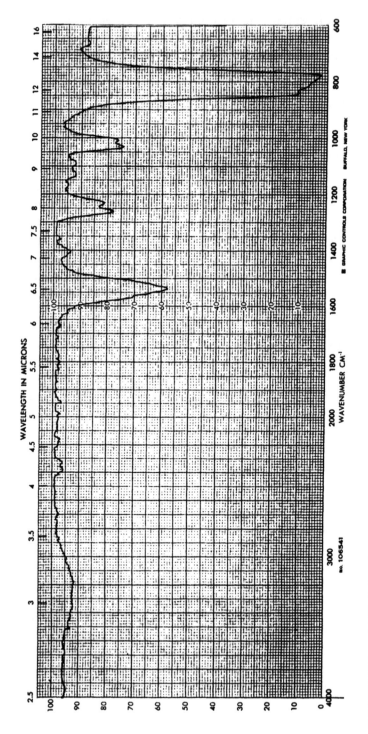

FIGURE 6.24   (a) The infrared spectrum of CCl_4. (b) The infrared spectrum of CH_2Cl_2.

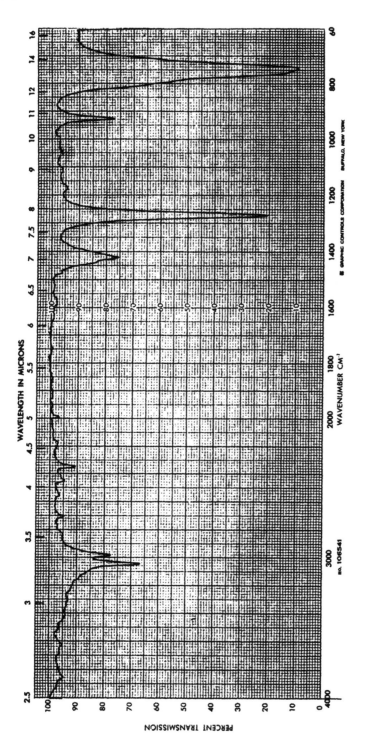

FIGURE 6.24b.

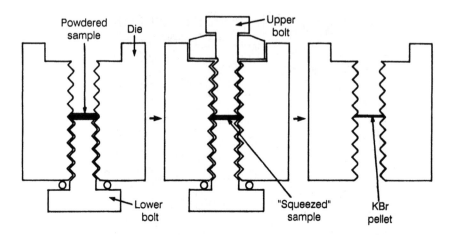

FIGURE 6.25   The procedure for making a KBr pellet with the use of a threaded pellet die.

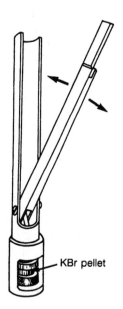

FIGURE 6.26   A pellet-making assembly which utilizes a laboratory hand press. (Reproduced from Spectra-Tech, Inc., Stamford, CT. With permission.)

displayed. The advantage of this method is that an analyte that does not form a good KBr pellet can be run with little or no problem.

6.4.4 Instrument Design

The traditional IR spectrometer is very similar to the UV/vis spectrophotometer, the so-called double-beam "dispersive" instrument utilizing the traditional slit/dispersing element/slit monochromator to create a narrow bandwidth, often referred to as the single wavelength, to shine on the sample. The modern IR instrument, however, utilizes an interferometer in which the light does not get dispersed, and yet the traditional IR spectrum is measured — It is called the Fourier Transform Infrared Spectrometer, FTIR. Both of these designs will now be discussed.

6.4.4a Double-Beam Dispersive

A diagram of the double-beam dispersive IR spectrometer is shown in Figure 6.28. It is very similar to the UV/vis spectrophotometer described in Section 6.3. There are some differences, however. The light source is typically a nichrome wire coil that has a high electrical resistance and emits intense heat (IR light) when an electric current is passed through it, much like the burner on an electric stove. The source may also be a "Globar," which is a silicon carbide rod, or a "Nernst Glower," which is a rare earth oxide cylinder. Both of these operate on basically the same principle as the nichrome wire. The detector is a thermocouple transducer, a device which converts heat energy into an electric signal.

The IR instrument is typically a "double-beam in space" instrument, whereas the UV/vis instrument is a "double-beam in time" instrument. "Double-beam in time" refers to the fact that the light beam from the source is split into two beams existing alternately in time. At one moment, the sample beam exists and passes through the sample compartment, while at the next moment, the reference beam exists and passes through the sample compartment. A rotating half-mirror (chopper) ahead of the

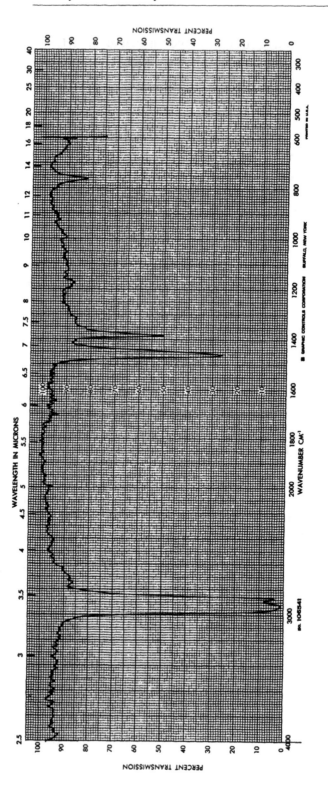

FIGURE 6.27   The infrared spectrum of mineral oil.

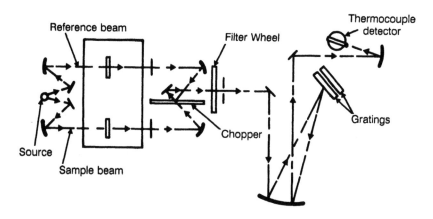

FIGURE 6.28   A diagram of a double-beam dispersive IR instrument. (Reproduced from Beckman Instruments, Inc., Fullerton, CA. With permission.)

sample compartment accomplishes this splitting of the beam (see Section 6.3). In the "double-beam in space" instrument, both light beams exist simultaneously. Two mirrors that point in different directions, as depicted in Figure 6.28, send two light beams through the sample compartment. Thus, both beams exist simultaneously and pass through the sample compartment together. The monochromator system, located between the sample compartment and detector in such an instrument, does utilize a chopper to combine the light beams, however, and the detection system is synchronized with the chopping frequency so as to differentiate the two signals and display the percent transmittance of the unknown sample at the recorder. As mentioned under liquid sampling (Section 6.4.2), placing the solvent of the analyte species in the reference path will cause all solvent absorption bands to disappear from the recorded spectrum just as the blank adjustment is automatically made in the UV/vis instruments discussed earlier. This does, however, require two exactly matched IR cells for accurate work. This cell matching is less important for qualitative work, which is a more common application of the technique.

The optical system for dispersive IR instruments has an additional requirement. The absorption of IR light by the mirrors, lenses, and dispersing element must be minimized. Thus, various reflection and transmittance gratings composed of nonabsorbing materials are used, often in combination with transmission filters, such as the "filter wheel" in Figure 6.28. Front-reflecting mirrors are used, and glass collimating lenses are absent. In addition, more than one grating are needed to accommodate the different regions of the IR spectrum, and this fact requires the instrument to stop scanning to allow the realignment of these gratings in the middle of a run and then to resume. Figure 6.28 shows two gratings. The gratings are usually made of glass or plastic and coated with aluminum.

6.4.4b Fourier Transform Infrared Spectrometry (FTIR)

The modern FTIR instrument is designed to perform the same functions as the dispersive instruments. Such an instrument, however, does not utilize a light dispersing monochromator and associated optics. The light from the source, typically the same type of source as in the dispersive instruments, passes through the optical path undispersed. The result, however, is the same — the infrared absorption spectrum, the plot of percent T vs wavelength, but it can be obtained much faster than with the dispersive instrument. We will undertake a simplified discussion as to how this happens.

In the FTIR instrument, the undispersed light beam passes through the sample, and all wavelengths and the corresponding absorption data are received at the detector simultaneously. A computerized mathematical manipulation known as the Fourier Transform is performed on this data in order to obtain absorption data for each individual wavelength. To present the data in a form that can utilize the Fourier Transform, the wave pattern created by combined constructive and destructive interference of all wavelengths of light from the source over time is utilized. This pattern, known as an interferogram, is created by moving one beam of light from the source through another. The device for doing this is known as an interferometer.

Figure 6.29 shows a diagram of an interferometer. It consists of a beam splitter and two mirrors, one fixed and one movable. Consider first a light source of a single wavelength. As light from the source strikes the beam splitter, the beam is split such that half of the intensity is transmitted to the movable mirror while half is reflected to the fixed mirror. Both beams are reflected back to the splitter where they join again and proceed toward the sample. If the distance from the splitter to the movable mirror is exactly equal to the distance to the fixed mirror, then their rejoining will result in constructive interference and the intensity reaching the sample will be a maximum. The position of the movable mirror, however, may also be such that complete destructive interference occurs. This would occur if the movable mirror distance is equal to the fixed mirror distance plus half of the wavelength. Other movable mirror distances would result

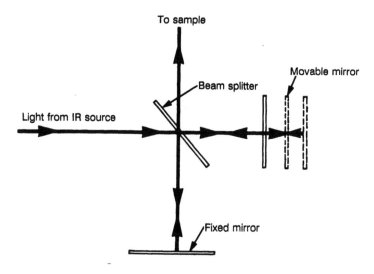

FIGURE 6.29   A diagram of an interferometer (see text for discussion).

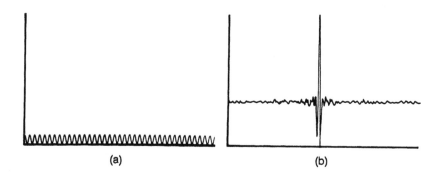

FIGURE 6.30   The two interferograms described in the text. (Reproduced from Chia, L. and Ricketts, S., *Basic Techniques and Experiments in Infrared and FT-IR Spectroscopy*, The Perkin-Elmer Corporation, Norwalk, CT, 1988. With permission.)

in intermediate intensities, the cycle would repeat as the distance is increased, and thus the "interferogram," known as a cosine wave, shown in Figure 6.30a would result. Now consider a light beam consisting of all wavelengths in the IR — in other words, the light from the light source in a typical IR instrument. In this case, the interferogram in Figure 6.30b would result. The strong intensity in the center occurs when the distances of both mirrors from the splitter are equal and we have complete constructive interference of all wavelengths. Motor-driving the movable mirror to small distances toward and away from the equidistance point "codes" the light and its wavelengths and creates the possibility for it to be "decoded" by the Fourier Transform mathematical manipulation by computer at the detector resulting in the IR spectrum of the sample.

The advantages of the FTIR over the dispersive technique are (1) it is faster, making it possible to be incorporated into chromatography schemes as we will see briefly in Chapters 9 and 10, and (2) the energy reaching the detector is much greater thus increasing the sensitivity.

6.4.5 Qualitative and Quantitative Analysis

The ultimate goal of IR analysis is, of course, the identification of the substance measured or the determination of the quantity of the substance measured. The usefulness of the technique lies mostly in the identification aspects, and we will emphasize this here. However, we will also briefly address the quantitative aspects as well.

6.4.5a Qualitative Analysis

Early on in this section (Section 6.4), we spoke of how specific bonds present in molecules give rise to absorption bands at specific corresponding wavelengths in the IR region. This is the basis of qualitative analysis using this technique, and we will now expand on this idea, especially discussing what wavelengths are absorbed by each type of bond.

The region of the electromagnetic spectrum involved here of course is the IR region. This spans the wavelength region from about 2.5 to about 17 μm, or, in terms of wave number, from about 4000 to about 600 cm^{-1}. IR spectra are transmission spectra calibrated most often in wave number, and so we will be speaking in terms of wave number for the remainder of this discussion and depicting the spectra with a 100% T baseline at the top of the chart and the peaks deflecting toward the 0% T level when an absorption occurs (see Figure 6.31 for an example). The absorption pattern derived from the particular molecule present in the path of the light is a molecular "fingerprint" just as it was in the UV/vis region (Section 6.3). The IR spectrum, however, is useful for an additional reason. The absorption bands, as they are typically recorded, appear much sharper and we have the ability to conclude that a molecule has a particular type of bond in its structure when we observe the corresponding characteristic absorption band in the spectrum. This is particularly true in the region from about 4000 to about 1500 cm^{-1} and less true in the 1500 to 600 cm^{-1} region. This latter region is thus often described separately as the "fingerprint" region and is used mostly to match, peak for peak, the spectrum of an unknown with a spectrum of a known, perhaps from a library of known spectra, such as the Sadtler library* (see Figure 6.32). Thus, we look for characteristic bands in the 4000 to 1500 cm^{-1} region to perhaps assign the unknown to a particular class of compounds, i.e., to narrow down the possible structures, and then look to the fingerprint region and the overall spectrum to make the final determination, matching peak for peak.

In addition to the location (i.e., wave number) of the absorption bands, it can also be useful to examine the width and depth of the absorption. The descriptions "broad" and "sharp" are often used to describe the width of the peaks, and "weak," "medium," and "strong" are used to describe the depth of the peaks. To understand these descriptions, refer to Figure 6.33. Spectra "a" and "c" show broad, strong peaks centered around 3300 cm^{-1}. Spectrum "b" shows a sharp, strong peak at 1685 cm^{-1}. Spectrum "c" shows a sharp, medium peak at 1220 cm^{-1} and a series of weak peaks between 1700 and 2000 cm^{-1}.

Table 6.2 correlates the location, width, and depth of various IR absorption patterns for some common kinds of bonds. By correlating Figure 6.33 with Table 6.2, it should be obvious that spectrum "a" in Figure 6.33 is that of an alcohol with no benzene rings, spectrum "b" is the spectrum of a compound containing a carbonyl group (such as an aldehyde) with no benzene ring, and spectrum "c" is the spectrum of an alcohol with a benzene ring. Figure 6.34 is a correlation chart showing the location of the absorption peaks of most bonds.

Finally, earlier the need for using a solvent such as carbon tetrachloride, dichloromethane, or nujol (mineral oil) was indicated for some applications. The analyst needs to know what the spectra of these solvents look like in order to be able to account for the interfering solvent peaks. Figures 6.24 and 6.27 gave the IR spectra for these compounds.

---

*   This refers to the collection of standard infrared spectra published by Sadtler Research Laboratories, Philadelphia, PA.

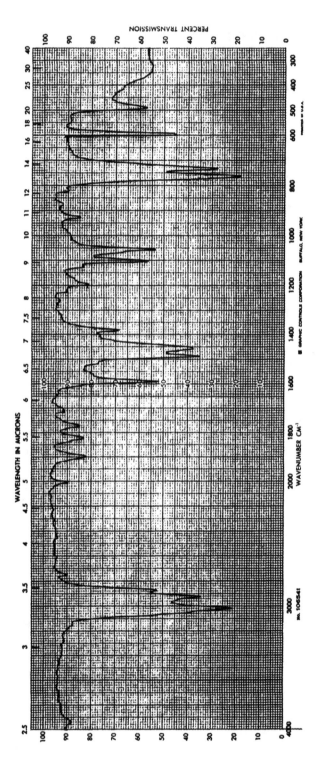

FIGURE 6.31   The infrared spectrum of toluene. Infrared spectra are transmission spectra with the peaks recorded from the top down as shown.

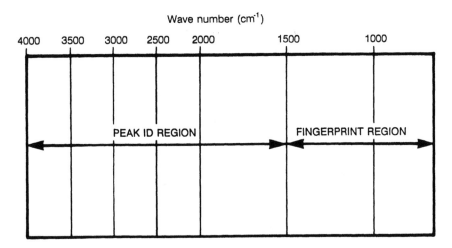

FIGURE 6.32   The "fingerprint" and "peak ID" regions of infrared spectra.

6.4.5b Quantitative Analysis

Quantitative analysis procedures using IR spectrometry utilize Beer's Law. Once the %T or absorbance measurements are made, the data reduction procedures are identical with those outlined previously in this chapter for UV/vis spectrophotometry. This means that the analyst could prepare just one standard solution to which to compare the unknown using the ratio and proportion scheme (Equation 6.10), determine the concentration by direct calculation (Equation 6.11), or prepare a series of standards and utilize a Beer's Law plot (see Figure 6.19 and accompanying discussion).

Reading the %T from the recorded IR spectrum for quantitative analysis can be a challenge. In the first place, there can be no interference from a nearby peak due to the solvent or other component. One must choose a peak to read that is at least nearly, if not completely, isolated.

Secondly, the baseline for the peak must be well defined. Since the baseline may not be completely straight across the wavelength region, the analyst does not endeavor to have the instrument trace across the 100%T line. Rather, something less than 100%T is chosen at the beginning of the run such that the baseline, even though it likely will not be straight, remains on scale (less than 100%T) for the entire scan. Thus, two %T readings must be taken, one for the baseline (corresponding to where the baseline would be if the peak were absent — a "blank" reading) and one for the minimum %T, the tip of the peak. The two readings are then

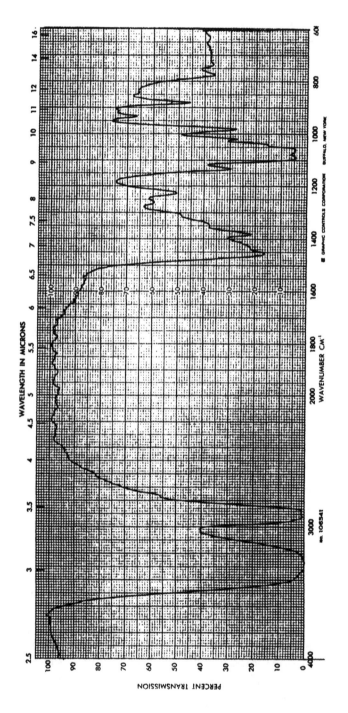

FIGURE 6.33   The IR spectra of (a) n-butyl alcohol, (b) isobutyraldehyde, and (c) benzyl alcohol (see text for discussion).

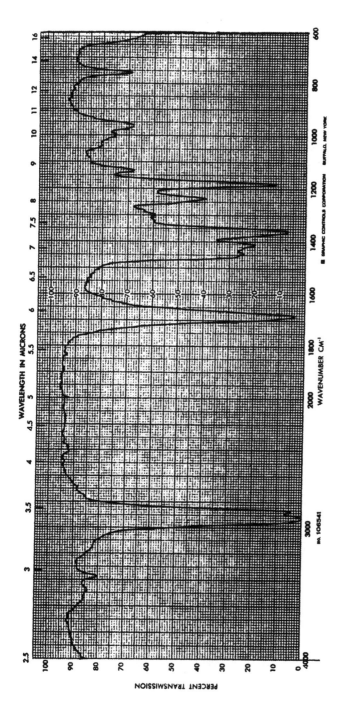

FIGURE 6.33b.

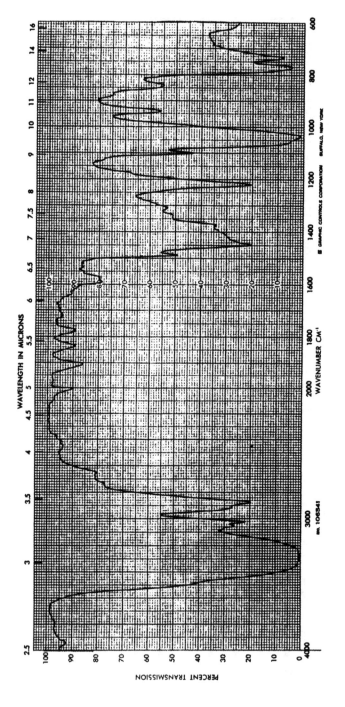

FIGURE 6.33c.

Table 6.2   Some Easily Recognizable IR Absorption Patterns

Bond	Description of Absorption Pattern
–C–H–, where C is not part of a benzene ring	Sharp, strong peak "on the low side of 3000 cm^{-1}," between about 2850 and 3000 cm^{-1}
–C–H–, where C is a part of a benzene ring or double bond	Sharp, medium peak "on the high side of 3000 cm^{-1}," between about 3000 and 3100 cm^{-1}
–C–H–, where C is a part of triple bond	Sharp, medium peak "on the high side of 3000 cm^{-1}," between about 3250 and 3350 cm^{-1}
–O–H–, in alcohols, phenols, and water, for example	Broad, strong peak centered at about 3300 cm^{-1}
–C=O, in aldehydes, ketones, etc.	Sharp, strong peak at about 1700 cm^{-1}
–C–C–, in benzene ring	Two sharp, strong peaks near 1500 and 1600 cm^{-1}, and a series of weak peaks (overtones) between 1600 and 2000 cm^{-1}, the latter in a case of a monosubstituted benzene ring

converted to absorbance, and the absorbance of the baseline is subtracted from the absorbance of the peak. The result is the absorbance of the sample. It is recommended, for accurate work, that two or more spectra of each standard and the unknown be recorded and an average of each absorbance calculated.

It should also be mentioned that the pathlength of the sample cell used must be constant for all standards and the unknown, or at least known so that a correction can be applied if necessary. Using care when filling and cleaning cells is also important to avoid alterations in pathlength due to excessive pressure or leaking.

6.5 FLUOROMETRY

Fluorometry is an analytical technique which utilizes the ability of some substances to exhibit fluorescence. Fluorescence is a phenomenon in which the substance appears to glow when a light shines on it. In other words, light of a wavelength different from the irradiating light is released or "emitted" following the absorption process. Most often the irradiating light is UV light, and the emitted light is vis light. The phenomenon is explained based on light absorption theory and what can happen to a

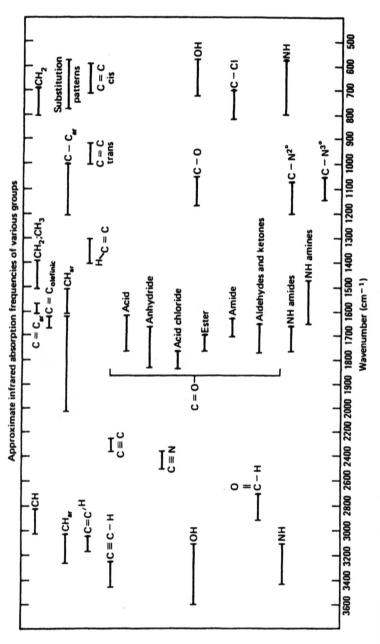

FIGURE 6.34   A correlation chart for IR spectroscopy. (Reprinted from Zubrick, J.W., *Organic Chemistry Survival Manual*. Copyright © John Wiley & Sons Inc., New York, 1988. With permission.)

chemical species in order to revert back to the ground state once the absorption — the elevation to an excited state — has taken place. Fluorescence can occur with both molecules and atoms. The present discussion will focus on molecules and complex ions. Atomic fluorescence will be discussed in Chapter 7.

All atoms and molecules seek to exist in their lowest possible energy state at all times. When a molecule is raised to an excited electronic energy state through the absorption of light, it is no longer in its lowest possible energy state and will seek to lose the energy it gained any way it can. Most often, the energy is lost through mechanical means, such as through collisions with other chemical species in the solution. However, there can be a direct jump back to the ground state with only some intermediate stops at some lower vibrational states in between. With such a jump back to the ground state, the energy the molecule gained as a result of the absorption process is lost in the form of light, and since it is light of less energy due to the accompanying small energy losses in the form of vibrational loss, the wavelength is longer. See Figure 6.35 for a graphical picture of this process.

The instrument for measuring fluorescence intensity for quantitative analysis is constructed with two monochromators, one to select the wavelength to be absorbed and one to select the fluorescence wavelength to be measured. In addition, the instrument components are configured so that the fluorescence measurement is optimized to be free of interference from transmitted light from the source.

This latter point means that the fluorescence monochromator and detector are not placed in a straight line with the source, absorption monochromator, and sample (such as in an absorption spectrophotometer), but are rather placed at a right angle as shown in Figure 6.36. Thus, there is a "right angle configuration" in a fluorometer to avoid any interference from the transmitted light from the light source.

Colored glass light filters are often used for the monochromators in fluorometers. Instruments with such filters are called filter fluorometers and are considerably less expensive than spectrophotofluorometers, which have standard slit/dispersing element/slit monochromators. Although these latter instruments are excellent for determining the proper wavelengths to be used and for precise work, filter fluorometers have proved quite satisfactory for most routine work and are much less expensive.

We have indicated that the intensity of the fluorescence emitted by the fluorescing species is proportional to the concentration of this species in solution. Fluorescence intensity is therefore the parameter to be measured

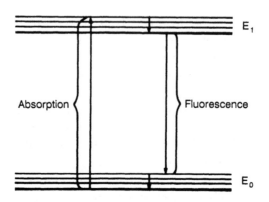

FIGURE 6.35   An energy level diagram showing the transitions occurring when absorption is accompanied by fluorescence.

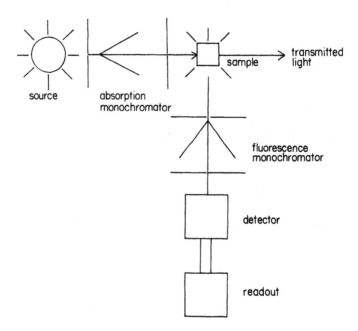

FIGURE 6.36   The basic fluorometer. The two monochromators can be glass filters. (From Kenkel, J., *Analytical Chemistry for Technicians*, Lewis Publishers, Inc., Chelsea, MI, 1988. With permission.)

and related to concentration. This means that the fluorescence intensity of a single standard solution, or a series of standards, is measured and related to concentration. A graph of fluorescence intensity vs concentration is expected to be linear in the concentration range studied. Proce-dures outlined in Chapter 5, Section 5.3, and also Section 6.3 for Beer's Law, are also applicable here.

The types of compounds that can be analyzed by fluorometry are limited. The drop of an electron to the ground state must be accompanied by the emission of light (it must be a direct drop). The kind of electron which is most apt to be able to do this is an electron, and, more specifi-cally, electrons found in benzene rings. Fused benzene ring systems, such as those in Figure 6.37, are especially highly fluorescent compounds. Metals can be analyzed by fluorometry if they are able to form complex ions by reaction with a ligand having the required electrons. An example is aluminum, as in Figure 6.38.

Fluorometry and absorption spectrophotometry are competing tech-niques in the sense that both are techniques for analyzing for molecular species. Each offers its own advantages and disadvantages, however. The number of chemical species that can exhibit fluorescence is very limited. However, for those species that do fluoresce, the fluorescence is generally very intense. Thus we can say that while absorption spectrophotometry is much more universally applicable, fluorometry suffers less from inter-ferences and is usually much more sensitive. Therefore, when an analyte does exhibit the somewhat rare quality of fluorescence, fluorometry is likely to be chosen for the analysis. The analysis of foods for vitamin content is an example, since vitamins such as riboflavin and niacin exhibit fluorescence and fluorometry would be relatively free of interference and would be very sensitive.

6.6 NUCLEAR MAGNETIC RESONANCE SPECTROSCOPY

6.6.1 Introduction

Throughout this chapter, we have studied instrumental analysis tech-niques which are based on the phenomenon of light absorption. We have discussed techniques utilizing light in the UV/vis region involving elec-tronic energy transitions — the elevation of electrons to higher energy states. We have also discussed techniques utilizing light in the IR region involving molecular vibrational and rotational energy transitions. In this

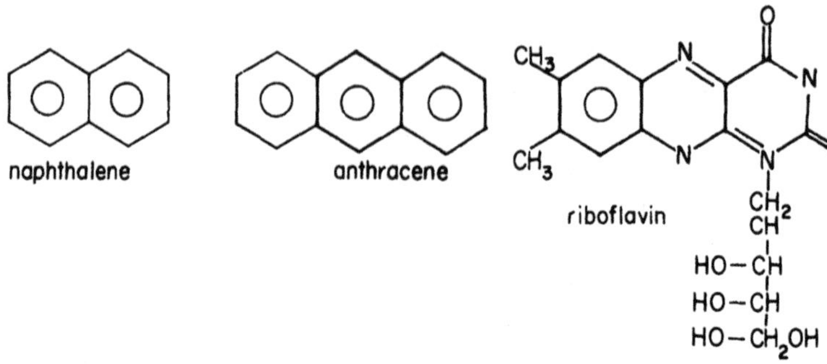

naphthalene   anthracene

riboflavin

FIGURE 6.37   Highly fluorescent compounds. (From Kenkel, J., *Analytical Chemistry for Technicians*, Lewis Publishers, Inc., Chelsea, MI, 1988. With permission.)

(highly fluorescent)

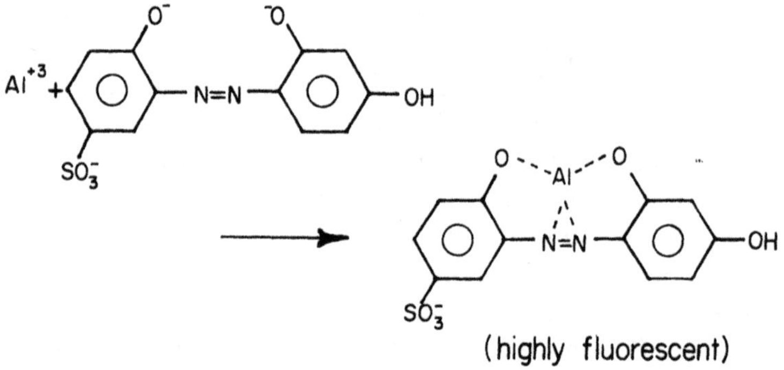

FIGURE 6.38   The formation of a complex ion of aluminum, which is highly fluorescent. (From Kenkel, J., *Analytical Chemistry for Technicians*, Lewis Publishers, Inc., Chelsea, MI, 1988. With permission.)

section, we introduce the concepts of Nuclear Magnetic Resonance Spectrometry (NMR). This technique utilizes light in the radio wave region of the spectrum involving nuclear spin energy transitions which occur in a magnetic field.

The absorption to be described is based on the theory and experimental evidence that the nuclei of the atoms bonded to each other in molecules spin on an axis like a top. Since any given nucleus is positively charged, a small magnetic field exists around it. If we were to bring a spinning nucleus into an external magnetic field, such as between the poles of a magnet, the nucleus, representing the smaller magnetic field, will align itself to the external field. It is possible to become aligned either in the same direction as the external field or in the opposite direction. Alignment

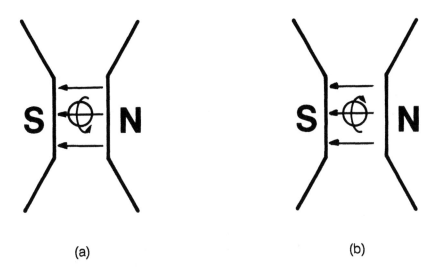

(a)   (b)

FIGURE 6.39   Alignment of a spinning nucleus (a) with a magnetic field and (b) opposed to a magnetic field.

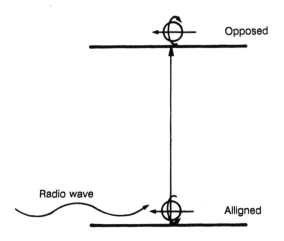

FIGURE 6.40   An energy level diagram showing the transition from one nuclear spin energy state to another.

in the opposite direction represents a slightly higher energy state (see Figure 6.39). The energy difference between the two different alignments is on the order of the radio wave wavelengths. Thus light in the radio wave region can be absorbed by molecules in a magnetic field so as to cause this nuclear spin energy transition (see Figure 6.40). While the phenomenon just described applies to nuclei of all elements, NMR has found its most useful application in the measurement of the hydrogen nucleus, probably because the vast majority of organic structures contain hydrogen atoms. For this reason, it is sometimes also referred to as Proton Magnetic Resonance. The application lies mostly in the determination of the structure of organic compounds; thus, it is mostly a qualitative analysis tool for such compounds.

6.6.2 The Traditional Instrument

Central to the instrumental design is a large magnet with the north and south poles facing each other as shown in Figure 6.39. There are several different types of magnets in common use, but stability and the ability to produce a precise magnetic field are common requirements. The ability to carefully vary the strength of the field between the two poles and the ability to output the magnitude of the field over time to a recorder are also requirements. A pair of coils positioned parallel to the magnet poles and connected to a "sweep generator" permits precise scanning of the magnetic field. Additionally, incorporated into the unit are a radio frequency (RF) transmitter capable of emitting a precise frequency, an RF receiver/detector for detecting absorption, and an x-y recorder to plot the output of the detector vs the applied magnetic field. The sample is held in a 5-mm outside diameter glass tube containing less than 0.5 mL of liquid, which in turn is held in a fixture called the "sample probe." A schematic diagram of the traditional NMR spectrometer is shown in Figure 6.41.

A precise radio frequency is emitted by the transmitter, so there is no component needed to act as a monochromator. However, the receiver/detector warrants additional comment. There are two designs for detectors. One utilizes a coil wrapped around the sample tube as the transmitter and a second coil arranged at right angles to the transmitter coil as the detector. This unique design will detect a signal only if absorption has taken place. The other design utilizes a single coil wrapped around the sample which, with the use of an appropriately designed electronic circuit, acts both as the transmitter and the receiver. The x-axis of the recorder is connected to the scanning mechanism as described previously, and thus with absorption plotted on the y-axis, the magnetic field strength is plotted on the x-axis. The result is a plot of absorption vs field strength, the so-called NMR spectrum. (See the discussion of chemical shift below for a more precise discussion of what is plotted on the x-axis.)

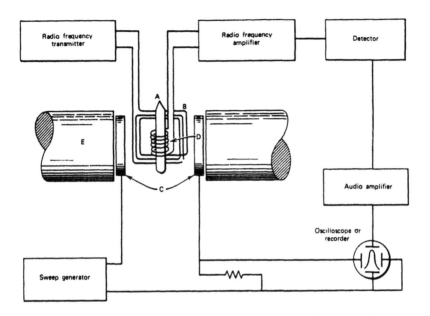

A = Sample Tube,   B = R-F Transmitter Coils,   C = Sweep Coils
D = Detector Coil,   E = Magnet.

FIGURE 6.41   Schematic diagram of an NMR spectrometer as described in the text. (Reprinted from Shriner, R.L., et al., *The Systematic Identification of Organic Compounds*, 6th ed., John Wiley & Sons, New York, 1980. With permission.)

The magnitude of the RF frequency used is variable. Some instruments are 60 MHz (megahertz, or one million cycles per seconds). Others are 100 MHz or higher. The magnitude of the frequency dictates the magnitude of the magnetic field strength required. The spread between the two energy levels in question (which corresponds to the energy of the RF frequency) depends on the strength of the field. A 60 MHz instrument requires a field strength of 14,092 gauss (the gauss is a unit of field strength), while a 100 MHz instrument requires 23,486 gauss.

6.6.3 Chemical Shifts

Electrons are, of course, present around and near the hydrogen nuclei in a molecule, and they are generating individual magnetic fields too. These very small fields oppose the applied field giving an effective applied field somewhat smaller than expected and therefore a slightly shifted absorption pattern, the so-called chemical shift. If all hydrogen nuclei in a molecule were "shielded" by electrons equally, they would all give an absorption peak at the same frequency/field combination. Due to the different environments surrounding the different hydrogens in an organic molecule, however, we find absorption peaks at locations representing each of these environments. In other words, each "type" of hydrogen in a structure would give a characteristic absorption peak at a specific field value. Methyl alcohol, CH_3OH, for example, would give two peaks, one representing the methyl hydrogens and one representing the hydroxyl hydrogen, while cyclohexane would give just one peak, etc. The importance of this information is that (1) from the number of different peaks, we can tell how many different kinds of hydrogen there are and (2) from the amount of shielding shown, we can determine the structure nearby.

Since the shifts can be extremely slight, making the actual field strength of the peak difficult to measure precisely, a reference compound, typically tetramethylsilane (TMS), is often added to the compound measured. All the hydrogens in the TMS structure are equivalent (Figure 6.42). It is then convenient to modify the x-axis to show the difference between the single TMS peak and the compound's peaks. Thus the x-axis typically represents a "difference factor" symbolized by the Greek letter lower case delta (δ), in which the single TMS peak is 0.0 ppm field strength difference and other peaks then are so many parts per million away from the TMS peak. Figure 6.43 shows such an NMR spectrum for methyl alcohol.

6.6.4 Peak Splitting and Integration

Additional qualitative information is possible with NMR spectra. First, a given absorption peak, while apparently arising from one particular kind of hydrogen, may appear to be split into two or more peaks, giving what are termed "doublets," "triplets," etc. Also, two obviously different kinds of hydrogens may give rise to just one peak (a singlet). Each of these phenomena is caused by effects of the immediate environment of the hydrogen and will present specific qualitative information. Second, the size of a peak is indicative of the number of hydrogens it represents. This,

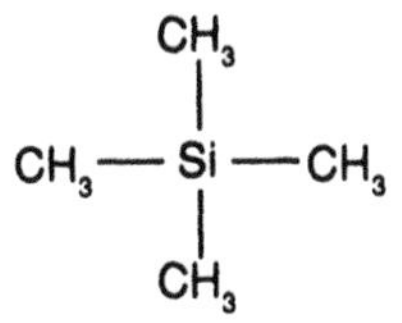

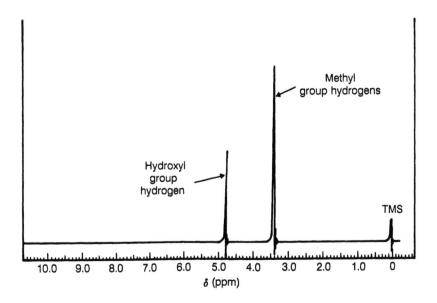

FIGURE 6.42   The structure of tetramethylsilane (TMS).

FIGURE 6.43   The NMR spectrum of methyl alcohol. (From Wade, L.G., Jr., *Organic Chemistry*, © 1987. Reprinted permission of Prentice-Hall, Englewood Cliffs, NJ.)

too, is important qualitative information, and so many NMR spectrometers are equipped with integrators for determining peak size. A second pen on the recorder is used to give an integration trace, which is then used to determine this number of hydrogens. Figure 6.44 is the spectrum of ethylbenzene, showing a quartet and triplet, as well as the integrator trace.

For more details on these and other concepts of NMR, please refer to comprehensive organic chemistry texts.

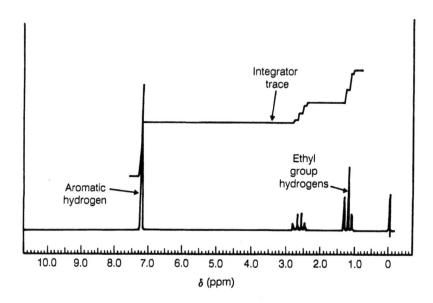

FIGURE 6.44   The NMR spectrum of ethylbenzene. (From Wade, L.G., Jr., *Organic Chemistry*, © 1987. Reprinted permission of Prentice-Hall, Inc., Englewood Cliffs, NJ.)

6.7 MASS SPECTROMETRY

6.7.1 Introduction

The final technique of molecular spectroscopy we will discuss is mass spectrometry. Unlike all the others in this chapter, this technique does not use light at all. Very briefly, the instrument known as a mass spectrometer utilizes a high energy electron beam to cause total destruction and fragmentation of the molecules of the sample. This fragmentation results in small charged "pieces" or fragments of the molecules which are then made to move through a magnetic field. The magnetic field affects each of the fragments differently according to their mass and charge, and thus they become separated. Finally, a detector sensitive to these fragments is placed in their path and, in combination with a readout device, such as a recorder or monitoring screen, allows the operator to determine the charge to mass ratio of each fragment and to identify not only the fragment, but the entire molecule. We will briefly discuss the details.

6.7.2 Instrument Design

There are two different instrument designs we will discuss. These are the magnetic sector mass spectrometer and the quadrupole mass spectrometer. Both designs consist of a system for sample introduction, the electron beam to create the fragmentation, a magnet to create the magnetic field, and a detection system. The entire path of the fragments, including the inlet system, must be evacuated from 10^{-4} to 10^{-8} torr. This requirement means that a sophisticated vacuum system must also be part of the setup. The reason for the vacuum is to avoid collisions of both the electron beam and the sample ions with contaminating particles which would alter the results.

The difference between the magnetic sector mass spectrometer and the quadrupole mass spectrometer lies in the design of the magnet. A diagram of a magnetic sector mass spectrometer is shown in Figure 6.45. In this instrument, the magnet is a powerful, variable field electromagnetic, the poles of which are shaped to cause a bending of the path of the fragments through a specific angle, such as 90°. It is possible to vary the field strength in such a way as to scan the magnetic field and to "focus" the ion fragments of variable mass to charge ratio onto the detector slit one at a time. In this way, specific fragments created at the electron beam can be separated from other fragments and detected individually.

A diagram of the quadrupole mass spectrometer is shown in Figure 6.46. Here, four short parallel metal rods with a diameter of about 0.5 cm each are utilized. These rods are aligned parallel to and surrounding the fragment path as shown. Two nonadjacent rods, such as those in the vertical plane, are connected to the positive pole of a variable power source, while the other two are connected to the negative pole. Thus, a variable electric field is created, and as the fragments enter the field and begin to pass down the center area, they deflect from their path. Varying the field creates the ability to "focus" the fragments one at a time onto the detector slit, as in the magnetic sector instruments. The quadrupole instrument is newer and more popular since it is much more compact and provides a faster scanning capability.

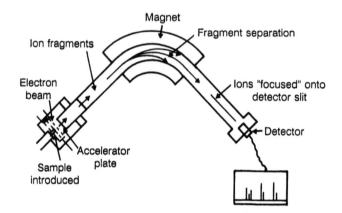

FIGURE 6.45   A magnetic sector mass spectrometer. (Adapted from Wade, L.G., Jr., *Organic Chemistry*, © 1987. Reprinted permission of Prentice-Hall, Englewood Cliffs, NJ.)

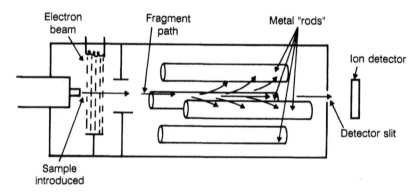

FIGURE 6.46   The quadrupole mass spectrometer.

6.7.3 Mass Spectra

The scanning of the magnetic field and the detection of fragments with a specific mass to charge ratio creates the possibility of manufacturing a plot of the fragment count vs mass to charge ratio. In other words, the fractions of all specific types of fragments (of particular mass and charge) resulting from a given sample can be determined. Such a plot is called the mass spectrum. An example is shown in Figure 6.47. Each vertical line evident in this figure represents a fragment of a particular mass to charge ratio given on the x-axis. The "intensity" of the lines represents the "count," or the number of fragments detected with that ratio. When

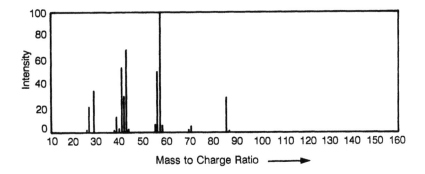

FIGURE 6.47   An example of a mass spectrum. (From Wade, L.G., Jr., *Organic Chemistry*, © 1987. Reprinted permission of Prentice-Hall, Englewood Cliffs, NJ.)

a certain molecule is introduced, it will be fragmented in a certain way and its characteristic mass spectrum will always be produced. One can see that it is a "molecular fingerprint," just as absorption spectra are molecular fingerprints, and that it is a powerful tool for identification purposes.

Modern mass spectrometer laboratories are linked to computers banks containing massive numbers of mass spectra obtained over the years. Specific identification, or at least a narrowing to specific possibilities, is often done by a computer that accesses these spectral files.

The mass spectrometer has been used as a detector in gas chromatography instruments and, more recently, in HPLC instruments. These combinations, referred to as GCMS and HPLC-MS, are discussed briefly in Chapters 9 and 10.

CHAPTER 7

ATOMIC SPECTROSCOPY

7.1 INTRODUCTION

This chapter discusses the techniques involving the absorption and emission of light by atoms (as opposed to molecules) and builds to some extent on the discussions of light, parameters of light, light absorption and light emission, and other concepts presented in Chapter 6. If these concepts are not well understood, we suggest that you first study the initial sections of Chapter 6 before beginning this chapter.

The analytical techniques known as atomic absorption (AA), flame photometry (FP), inductively coupled plasma (ICP), atomic fluorescence, and atomic emission spectrography are all included under the heading of "atomic" techniques or "Atomic Spectroscopy." Since the chemical species that absorb or emit light in these cases are atoms, the techniques are limited to sample systems from which atoms can be generated, especially including solutions of metal ions and excluding solutions of molecular species. Techniques for molecular species come under the heading of "Molecular Spectroscopy" and are discussed in Chapter 6. Atomic spectroscopy has been applied to a wide range of metals and also to some nonmetals. In this chapter, we will describe the theory, instrumentation, and application of each of the above listed techniques.

7.2 ATOMIZATION

First, how are metal atoms generated from metal ions? There are a number of methods. The earliest discovered method was with the use of a flame. When solutions of metal ions are placed in a flame, the solvent evaporates leaving behind crystals of the formerly dissolved salt. Dissociation into atoms then occurs; the metal ions "atomize" or are transformed into atoms. Flames are used for this purpose in most atomic absorption instruments and in all flame photometry and atomic fluorescence instruments. Such instruments, especially the AA, are easily recognized because of the centralized, hooded area in which a large flame, often 6 in. wide by 6 in. or more high, is located. All "atomizers," including the flame, are similar energy sources. Some are of the "atomic vapor" generator variety. Examples of these include the graphite furnace, the Delves cup, and the borohydride vapor generator (see Section 7.6). Another type uses an inductively coupled plasma (Section 7.7), and another uses a spark or arc across a pair of electrodes (Section 7.8). In each case, in addition to atomization, excitation of the atoms also occurs. The concept of excitation will be discussed in Section 7.3. In the remaining part of this section, we discuss the details of the flame atomizer; others will be discussed in those later sections.

7.2.1 Fuels and Oxidants

All flames require both a fuel and an oxidant in order to exist. Bunsen burners and Meker (Fisher) burners utilize natural gas for the fuel and air for the oxidant. The temperature of such a flame is 1800 K maximum. In order to atomize and excite most metal ions and achieve significant sensitivity for quantitative analysis by atomic spectroscopy, however, a hotter flame is desirable. Most AA and FP flames today are air-acetylene flames — acetylene for the fuel, air for the oxidant. A maximum temperature of 2300 K is achieved in such a flame. Ideally, pure oxygen with acetylene would produce the highest temperature (3100 K), but such a flame suffers from the disadvantage of a high burning velocity, which decreases the completeness of the atomization and therefore lowers the sensitivity. Nitrous oxide (N_2O) used as the oxidant, however, produces a higher flame temperature (2900 K), while burning at a low rate. Thus,

Table 7.1   A Listing Showing Which Oxidant, Air or Nitrous Oxide, is
Recommended for the Various Metals and Nonmetals Analyzed
by AA

Air:
   Lithium, Sodium, Magnesium, Potassium, Calcium, Chromium, Manganese,
   Iron, Cobalt, Nickel, Copper, Zinc, Arsenic, Selenium, Rubidium, Ruthenium,
   Rhodium, Palladium, Silver, Cadmium, Indium, Antimony, Tellurium,
   Cesium, Iridium, Platinum, Gold, Mercury, Thallium, Lead, Bismuth

Nitrous Oxide:
   Beryllium, Boron, Aluminum, Silicon, Phosphorus, Scandium, Titanium,
   Vanadium, Gallium, Germanium, Strontium, Yttrium, Zirconium, Niobium,
   Molybdenum, Tin, Barium, Lanthanum, Hafnium, Tantalum, Tungsten,
   Rhenium, Praseodymium, Neodymium, Samarium, Europium, Gadolinium,
   Terbium, Dysprosium, Holmium, Erbium, Thulium, Ytterbium, Uranium

From Perkin-Elmer Corporation, Norwalk, CT; product literature.

N_2O-acetylene flames are fairly popular. The choice is made based on
which flame temperature/burning velocity combination works best with
a given element. Since all elements have been studied extensively, the
recommendations for any given element are available. Table 7.1 lists most
metals and the recommended flame for each. Air-acetylene flames are
obviously the most commonly used.

7.2.2 Burner Designs

There are two designs of burners for the flame atomizer that are in
common use. These are the so-called "total consumption burner" and the
"premix burner". In the total consumption burner (Figure 7.1), the fuel,
oxidant, and sample all meet for the first time at the base of the flame.
The fuel (usually acetylene) and oxidant (usually air) are forced, under
pressure, into the flame, whereas the sample is drawn by aspiration into
the flame through a small diameter plastic tube. The rush of the fuel and
oxidant through the burner head creates a vacuum in the sample line and
draws the sample from the sample container into the flame. This type
of burner head is used in flame photometry and is not useful for atomic
absorption. The reason for this is that the resulting flame is turbulent and
nonhomogeneous — a property that negates its usefulness in AA, since
the flame must be homogeneous for the same reason that different sample
cuvettes in molecular spectroscopy must be closely matched. One would

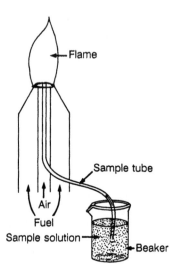

FIGURE 7.1   A diagram of a total consumption burner.

not want the absorption properties to change from one moment to the next because of the lack of homogeneity in the flame. The lack of homogeneity, however, does not affect the quality of the data obtained with a flame photometer. The reason will become clear in Section 7.4.

The premix burner does away with the homogeneity difficulty and is the burner typically used in flame AA. The sample is again drawn from the sample container by aspiration through a small diameter flexible plastic tube, nebulized, (split into a fine mist) and mixed with the fuel and oxidant with the use of a flow spoiler (such as a set of baffles) prior to introduction into the flame. The burner head typically used is rectangular and has a 4- to 6-in. long slot through which the premixed fuel, oxidant, and sample emerge and are ignited, creating the flame. Figure 7.2 is a diagram of this design. In the nebulizer, the sample enters an even smaller diameter tube and then impacts a glass bead, creating the fine mist. There is an adjustment on the nebulizer which controls the aspiration rate and thus the amount of sample reaching the flame. This adjustment is usually set so as to obtain the maximum absorbance on the readout. Most instruments are equipped to accept a variety of fuels and oxidants. As the gas combinations are varied, it is usually necessary to change the burner head to one suitable for the particular combination chosen. A faster burning mixture would require a burner head with a smaller slot so as to discourage drawing the flame inside the burner head causing a flashback.

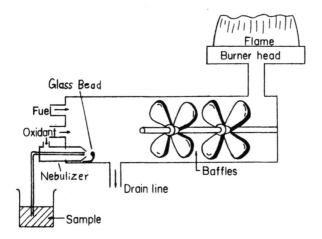

FIGURE 7.2   A diagram of a Premix Burner. (From Kenkel, J., *Analytical Chemistry for Technicians*, Lewis Publishers, Inc., Chelsea, MI, 1988. With permission.)

Flashbacks can also occur when air is drawn back through the drain line illustrated at the bottom of the premix chamber in Figure 7.2. The drain line is necessary to allow droplets of solution that do not make it to the flame to drain out. This problem is solved by forming a trap in the drain line (a flexible plastic tube) and keeping the end of the drain immersed in the waste solution contained in a bottle below the instrument.

7.3 EXCITATION

Following atomization, as indicated in the previous section, excitation of the atoms occurs to a small extent in the flame atomizer. A very small percentage of the atoms absorb energy from the flame and are elevated to an excited state. The technique of flame photometry is derived from this process as we will see. The remaining high percentage of ground state atoms that are therefore also present in the flame are subject to excitation with a light source. Flame atomic absorption is derived from this process.

As discussed in Chapter 6, basic atomic theory holds that electrons in atoms exist in energy levels around the nucleus. This theory also holds that electrons can be moved from one energy level to a higher one if conditions are right. These conditions consist of (1) the absorption of sufficient energy by the electrons and (2) a vacancy for the electron with this greater energy in a higher level. In other words, if an electron absorbs the energy required to be promoted to a higher vacant energy level, then it will be promoted to that level, and the atom will have undergone a transition from the ground electronic state to an excited electronic state. These "electronic energy levels" are the only ones that exist in atoms (no vibrational levels as with molecules), and thus electronic "transitions" are the only kind of energy transitions that can occur. This fact accounts for many of the differences between molecular spectrophotometers and atomic spectrophotometers and between the theories associated with each.

Since all energy transitions that take place in atoms are purely electronic, only individual, discrete, electronic energy transitions are possible. These transitions involve the elevation of electrons from one electronic level to another, as depicted in Figure 7.3a. Since no vibrational transitions take place, only a limited number of energies, those corresponding to the very specific electronic transitions, have a chance of being absorbed. For flame atomic absorption, in which wavelengths of light get absorbed by the atoms, the result is a "line" absorption spectrum (Figure 7.3b) rather than a "continuous" absorption spectrum as found for molecules. The term "line" is used here in reference to the individual lines, or wavelengths, of absorption evident in Figure 7.3b.

Those atoms excited by the light source in flame atomic absorption are those that are measured by this technique. Those atoms excited by the flame are those that are measured by flame photometry. Details of these two techniques are given in the following two sections. As mentioned earlier, atomization can be caused by methods other than a flame, and thus excitation can also be caused by energy sources other than a flame or light source. Details of techniques associated with these are given in Sections 7.6, 7.7, and 7.8.

7.4 FLAME PHOTOMETRY

7.4.1 Introduction

Flame photometry is the technique which measures the atoms excited by a flame (and not by a light source). This measurement is possible because atoms that find themselves in an excited state (such as those

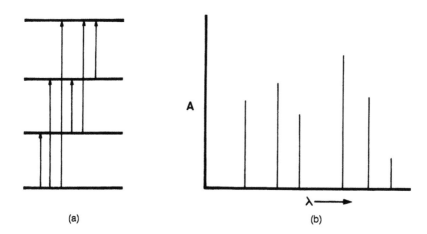

(a)   (b)

FIGURE 7.3   (a) A hypothetical energy level diagram showing four electronic levels
                 and the transitions that are possible within these levels. (b) A line
                 absorbtion spectrum. Each transition indicated in (a) corresponds to
                 a line in the spectrum in (b).

found naturally from a solution aspirated into a flame) will readily lose
the gained energy in order to revert back to the ground state. Since the
magnitude of the energy lost is on the order of light energy, light is
emitted. The wavelengths of the emitted light correspond to those same
wavelengths as those that were absorbed in the flame atomic absorption
technique discussed briefly in the last section, since exactly the same
energy transitions occur, except in reverse. Figure 7.4 illustrates this
phenomenon. The spectrum shown in Figure 7.4b is a "line emission"
spectrum. Thus a line spectrum can be either an absorption spectrum or
an emission spectrum depending on the process measured.
     Each individual metal has its own characteristic emission and absorption
pattern — its own unique set of wavelengths emitted or absorbed and
its own unique line emission or absorption spectrum. This is because each
individual metal atom has its own unique set of electronic levels. This
fact is demonstrated in a simple laboratory test known as the "flame test."
Sodium atoms present in a simple low temperature Bunsen burner flame
will emit a characteristic yellow light. Potassium atoms present in such
a flame will emit a violet light. Lithium and strontium atoms emit a red
light. The transitions occurring in the sodium atoms are such that the line
spectrum that is emitted corresponds to yellow light, while those occurring
in the potassium atom correspond to violet light, etc. (see Figure 7.5). The
usefulness of emission spectroscopy for qualitative analysis is thus apparent, especially given the fact that we can utilize monochromators and

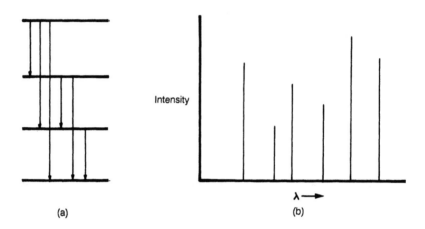

FIGURE 7.4   (a) A hypothetical set of transitions from higher electronic states to lower electronic states. (b) The line emission spectrum that results from the transitions in (a).

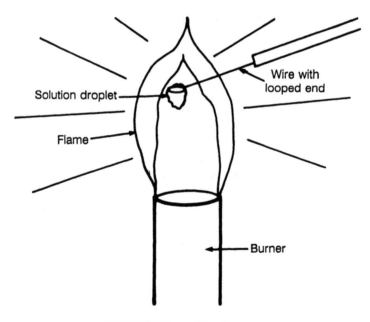

FIGURE 7.5   The flame test.

detectors to precisely measure the wavelengths involved. Indeed, all emission spectroscopy instruments, including flame photometers, ICPs, and emission spectrographs, are useful for qualitative analysis. Flame photometry and ICP, however, find their greatest application in quantitative analysis, as we will see.

7.4.2 Instrumentation

The simple flame photometer consists of three parts: the flame, a monochromator (Chapter 6, Section 6.3), and a detector/readout (Chapter 6, Section 6.3). As mentioned earlier, the flame originates from the total consumption burner. Of course, a premix burner may also be used, and indeed instruments, i.e., flame atomic absorption instruments, that utilize the premix burner as a matter of necessity, can also function as flame photometers, as we will discuss shortly. Instruments, however, that are manufactured and sold as flame photometers cannot be used as flame AA instruments, since they are designed with a total consumption burner and no light source to excite the atoms in the flame. Why is the total consumption burner a satisfactory burner for FP? While the flame from this burner is not homogeneous in its makeup, the intensity of the light emitted by the atoms in the flame is not affected. Such emission is normally quite stable. It is, of course, the intensity of this emission that is measured for quantitative analysis, as we will see in the next section.

The second major component is the monochromator. In molecular absorption instruments (Chapter 6, Section 6.3), a monochromator is needed to isolate the wavelength of maximum absorbance from all other wavelengths coming from the light source. In flame photometry, there is no light source (independent of the flame itself), and thus the monochromator is needed for a different reason. For quantitative analysis, the intensity of the emitted light must be measured. For maximum sensitivity, we want to measure the intensity of the most intense wavelength emitted by the flame. The monochromator is thus tuned to this wavelength. Which wavelength is it? The flame emits the line spectrum of the element to be measured. The most intense line in the spectrum is generally the wavelength to zero in on. It is possible, however, that this line is not the optimum line due to interferences. If this is the case, then a different wavelength may be optimum. "Primary" and "secondary" lines, etc. are thus often defined. Generally, a flame photometry procedure has been well researched and the wavelength suggested in such a procedure will be the best wavelength.

The monochromator is useful for another reason. In the molecular spectroscopy techniques discussed in Chapter 6, the sample cuvette is located in a light tight box (sample "compartment"), such that room light is not a problem. With flame techniques, however, the flame must be in an open area of the instrument. The monochromator thus also serves to isolate the desired wavelength from room light.

The final component of a flame photometer is the detector/readout and associated electronics. The detector is essentially the same component described in Chapter 6, Section 6.3. It is a device that generates an amplified electronic signal when light strikes it. In this case, however, the associated electronics is designed to measure the intensity of the light, rather than a transmittance or absorbance. Thus, the readout displays relative intensity. Figure 7.6 shows a schematic diagram of the flame photometer.

7.4.3 Application

Application of flame photometry can be of both a qualitative and quantitative nature. For qualitative analysis, the determination of which wavelengths are emitted is the key, since each element emits its own characteristics line spectrum. Thus, one would check for the emission of an element's wavelengths by tuning the monochromator to the expected wavelengths for the element in question and checking for a readout of intensity. The expected wavelengths can be obtained from tables of emission lines such as Table 7.2.

Quantitative analysis by flame photometry is encountered more frequently than qualitative analysis in an analytical laboratory, however. As implied previously, the intensity of the emitted light is directly and linearly proportional to the concentration. Thus the standard curve for quantitative analysis is a plot of intensity vs concentration (Figure 7.7). As discussed in this book for other techniques, the standard curve is prepared by measuring the readout for a series of standard solutions of the element of interest. Table 7.2 also gives the detection limit and application of the elements typically determined by flame photometry.

It should be pointed out that other atomic techniques, especially flame and graphite furnace AA and ICP, have distinct advantages over flame photometry in many respects, especially sensitivity and linear concentration ranges. However, the flame photometry technique has been around longer

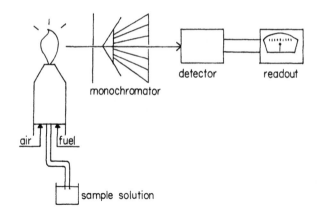

FIGURE 7.6   A schematic diagram of a flame photometer. (From Kenkel, J., *Analytical Chemistry for Technicians*, Lewis Publishers, Inc., Chelsea, MI, 1988. With permission.)

Table 7.2   The Elements Typically Determined by Flame Photometry, Their Corresponding Primary Lines, Their Detection Limit, and Applications

Element	Primary Line (nm)	Detection Limit (ppm)	Applications
Lithium	670.8	0.00003	Water, wastewater, biological fuids, soil extracts, etc.
Potassium	766.5	0.0005	Water, wastewater, biological fluids, soil extracts, etc.
Sodium	589.0	0.0005	Water, wastewater, biological fluids, soil extracts, etc.
Strontium	460.7	0.09	Water, wastewater, soil extracts, etc.

and some routine analyses are well established, especially in clinical laboratories. Also, the instruments are much less expensive.

The flame photometer also has a specific application as a detector for gas chromatography (see Chapter 9, Section 9.7).

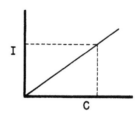

FIGURE 7.7   The standard curve for quantitative analysis in flame photometry.

7.5 FLAME ATOMIC ABSORPTION

7.5.1 Introduction

Of the various atomization techniques (flame, graphite furnace, Delves cup, and borohydride vapor generator) with which we combine the use of a light source for excitation producing an "absorption" technique, the flame atomizer is the oldest and the most common. We refer to these techniques collectively as atomic absorption (AA) techniques, but the one which uses the flame atomizer is also often referred to as simply atomic absorption. In order to make a distinction, however, between the use of the flame and the others, we refer to the absorption technique which utilizes a flame as the atomizer as flame atomic absorption. This is the technique discussed in this section. The others will be discussed in Section 7.6.

As stated previously, only a very small percentage of the atoms in the flame are actually present in an excited state at any given instant. The exact percentage depends on the flame temperature, but at the hottest temperature of any flame atomizer used for AA (2900 K), the proportion of atoms actually in the excited state at a given instant is much less than 0.01% of the total. Thus there is a large percentage of atoms that are in the ground state and available to be excited by some other means, such as a beam of light from a light source. Flame AA takes advantage of this fact and uses a light beam to excite these ground state atoms in the flame. Thus AA is very much like molecular absorption spectroscopy in that light absorption (by these ground state atoms) is measured and related to concentration. The major differences lie in instrument design, especially with respect to the light source, the sample "container," and the placement of the monochromator.

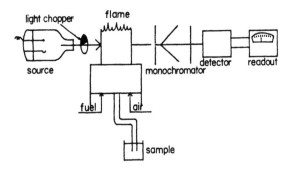

FIGURE 7.8   The Basic Flame AA Instrument. (From Kenkel, J., *Analytical Chemistry for Technicians*, Lewis Publishers, Inc., Chelsea, MI, 1988. With permission.)

7.5.2 Instrumentation

The basic flame AA instrument is shown in Figure 7.8. The light source, in most cases a hollow cathode tube, is a lamp that emits exactly the wavelength required for the analysis (without the use of a monochromator). This light beam is directed at the flame containing the sample. The flame standing atop the slot of the premix burner (Section 7.2) is wide (the width of the burner head slot). This width represents the pathlength for Beer's Law (Chapter 6, Section 6.3) considerations. The 4–6 in. width thus aids in determining small concentrations of the metal being analyzed. The light beam then enters the monochromator, which is tuned to the recommended line, the so-called "primary" line, from the metal's line spectrum. This line emerges from an adjustable slit opening and is thus the wavelength that strikes the detector. Since this is an absorption technique, the electronics associated with the detector/readout is designed to display either absorbance or transmittance on the readout.

7.5.2a The Hollow Cathode Lamp

How is it that the hollow cathode lamp emits exactly the wavelength required without the use of a monochromator? The reason is that atoms of the metal to be tested are present within the lamp, and when the lamp is on, these atoms are supplied with energy which causes them to elevate to the excited states. Upon returning to the ground state, exactly the same wavelengths that are useful in the analysis are emitted, since it is the analyzed metal with exactly the same energy levels that undergoes excitation. Figure 7.9 is an illustration of this. Therefore, the hollow cathode lamp must contain the element being determined. A typical atomic absorption laboratory has a number of different lamps in stock which can be interchanged in the instrument, depending on what metal is being determined. Some lamps are "multielement," which means that several different specified kinds of atoms are present in the lamp and are excited when the lamp is on. The light emitted by such a lamp consists of the line spectra of all the kinds of atoms present. No interference will usually occur, however, since the monochromator isolates a wavelength of our own choosing.

The exact mechanism of the excitation process in the hollow cathode lamp is of interest. Figure 7.10 is a closeup view of this lamp and of the mechanism. The lamp itself is a sealed glass envelope containing either argon or neon gas (neon shown in figure). When the lamp is on, neon atoms are ionized, with the electrons drawn to the anode (+ charged electrode), while the neon ions (Ne^+) "bombard" the surface of the cathode (– charged electrode). The metal atoms, M, in the cathode are sputtered from the surface and raised to the excited state as a result of the bombardment. When the atoms return to the ground state, the characteristic line spectrum of that atom is emitted. It is this light which is directed at the flame, where unexcited atoms of the same element absorb the radiation and are themselves raised to the excited state. As indicated previously, the absorbance is measured and related to concentration.

7.5.2b Electrodeless Discharge Lamp

A light source known as the electrodeless discharge lamp (EDL) is sometimes used. In this lamp, there is no anode or cathode. Rather, a small, sealed quartz tube containing the metal or metal salt and some argon at low pressure is wrapped with a coil for the purpose of creating a radio frequency (RF) field. The tube is thus inductively coupled to an RF field and the coupled energy ionizes the argon. The generated electrons collide with the metal atoms, raising them to the excited state. The characteristic line spectrum of the metal is thus generated and is directed at the flame just as with the hollow cathode tube. EDLs are available for

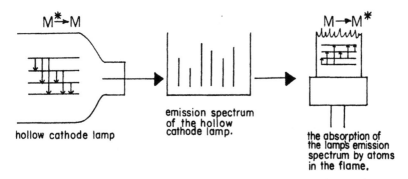

FIGURE 7.9   The light emitted by the hollow cathode lamp is exactly the wavelength needed by the atoms in the flame. (From Kenkel, J., *Analytical Chemistry for Technicians*, Lewis Publishers, Inc., Chelsea, MI, 1988. With permission.)

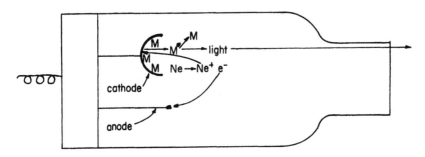

FIGURE 7.10   The hollow cathode lamp and the process of metal atom excitation and light emission. (From Kenkel, J., *Analytical Chemistry for Technicians*, Lewis Publishers, Inc., Chelsea, MI, 1988. With permission.)

17 different elements. Their advantage lies in the fact that they are capable of producing a much more intense spectrum and thus are useful for those elements whose hollow cathode lamps can produce only a weak spectrum.

7.5.2c Light Chopper

Another item relating to instrument design that is different from the molecular absorption spectrophotometers is the fact that there are two sources of light that enter the monochromator and strike the detector. How can the detector measure only the intensity of the light that does not get absorbed and not measure the light emitted by the flame, since both sources of light are of the same wavelength? Notice in Figure 7.8 that there is a light chopper placed between the lamp and the flame. The light is "chopped" with a rotating half-mirror, a component similar to the beam splitting device discussed for double-beam instruments in Chapter 6. The detectors thus sees alternating light intensities. At one moment, only the light emitted by the flame is read, since the light from the lamp is cutoff, while at the next moment, the light from both the flame emission and the transmission of the lamp's light is measured, since the lamp's light is allowed to pass. The electronics of the detector is such that the emission signal is subtracted form the total signal and this difference then is what is measured. Absorbance is usually displayed on the readout.

7.5.2d Double-Beam AA

There is also a difference in the design of double-beam instruments. In flame AA, the second beam does not pass through a second sample container as it does for the molecular instruments, since a second flame would be required and it would need to be identical to the sample flame. Rather, the second beam simply bypasses the flame and is relayed to the detector directly. This design eliminates variations due to fluctuations in source intensity (the major objective of a double-beam), but does not eliminate effects due to the flame or other components in the sample (blank components). These must still be adjusted for by reading the blank at a separate time. Also, the flame's emission continues to be accounted for via the chopper as with the single-beam instrument (see Figure 7.11).

7.5.2e Monochromator, Detector, and Readout

Descriptions of the monochromator and detector/readout are similar to that presented for molecular instruments in Chapter 6. One important difference is the placement of the monochromator. It is situated between the light source and the sample in UV/vis molecular instruments, but between the sample and the detector in atomic instruments, an arrangement similar to IR molecular instruments. In AA instruments, the light source (the hollow cathode lamp) does not require a monochromator in order

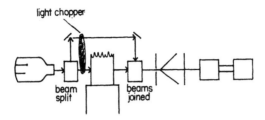

light chopper

beam split

beams joined

FIGURE 7.11   A representation of a double-beam AA instrument.

to obtain the desired wavelength, as indicated previously. However, once the light has passed through the flame, which, as with flame photometry, is located in an open, exposed portion of the instrument, it is desirable to screen out light from the flame and light from the room, as well as extraneous lines from the source before the light strikes the detector. Thus the monochromator is placed just before the detector. Another difference between the AA monochromator and the monochromator in molecular instruments is that there is a manually adjustable exit slit opening on the AA instrument. One can thus widen or narrow the slit manually to optimize the optics for the metal being analyzed.

The readout is in the form of a meter display on the instrument, a recorder, or through data acquisition with a computer. In a typical experiment, the readout consists of a series of absorbance readings with blank readings in between, since the blank is ordinarily read before each sample or standard. A strip-chart recording, for example, of the absorbance values for a series of five standard solutions would appear as in Figure 7.12.

7.5.3 Application

Quantitative analysis in atomic absorption spectroscopy utilizes Beer's Law. The standard curve is a Beer's Law plot, a plot of absorbance vs concentration. The usual procedure, as with other quantitative instrumental methods, is to prepare a series of standard solutions over a concentration range suitable for the samples being analyzed, i.e., such that the expected sample concentrations are within the range established by the standards. The standards and the samples are then aspirated into the flame, and the

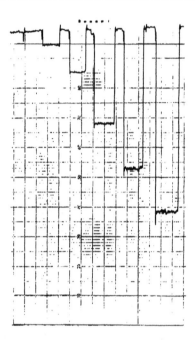

FIGURE 7.12   Strip-chart recording of the absorbance values of a series of standard solutions measured by an AA instrument. (From Kenkel, J., *Analytical Chemistry for Technicians*, Lewis Publishers, Inc., Chelsea, MI, 1988. With permission.)

absorbances read from the instrument. The Beer's Law plot will reveal the useful linear range and the concentrations of the sample solutions. In addition, information on useful linear ranges is often available for individual elements and instrument conditions from manuals of analytical methods available from instrument manufacturers, such as the Perkin-Elmer Corporation, Norwalk, CT.

7.5.3a Interferences

Interferences can be a problem in the application of AA. Interferences can be caused by chemical sources (chemical components present in the sample matrix) or instrumental sources (so-called "spectral" interferences arising from an instrument condition that turns out to be less than optimal for a particular sample). Chemical interferences usually result from incomplete atomization caused by an unusually strong ionic bond. An example is in the analysis of a sample for calcium. The presence of sulfate or phosphate in the sample matrix along with the calcium suppresses the reading for calcium because of incomplete atomization due to the strong ionic bond between calcium and the sulfate and phosphate. This results in a low reading for the calcium in the sample in which this interference exists. The usual solution to this problem is to add a substance to the sample which would chemically free the element being analyzed, calcium in our example, from the interference. With our calcium example, the substance that accomplishes this is lanthanum. Lanthanum sulfates and phosphates are more stable than the corresponding calcium salts, and thus the calcium is free to atomize when lanthanum is present.

In addition to the above method for removing a chemical interference, another possible solution may be to exactly match the matrix of the standard and blank to that of the sample, and thus the interference would be present in all solutions tested (at the same concentration) and this would negate the problem, although sensitivity may be decreased due to a smaller concentration of metal ions that get atomized. The method of standard additions described in Chapter 5 is a way in which this may be accomplished.

An example of a spectral interference is when the spectral line of the element being determined nearly overlaps the line of another element in the sample, such that some of the light from the hollow cathode lamp will be absorbed by this interfering element, creating an absorbance reading that is high. We call this an instrumental interference because the slit opening setting suggested by the instrument manufacturer is too wide to totally isolate the desired wavelength emerging from the monochromator. The solution is to use a narrower slit width or to zero in on a different line, a so-called "secondary" line for the analyte element rather than the "primary" line. Recommended secondary lines are also found in instrument manufacturer's methods manuals and other literature sources.

7.5.3b Safety and Maintenance

There are a number of important safety considerations regarding the use of AA equipment. These center around the use of highly flammable acetylene, as well as the use of a large flame, and the possible contamination of laboratory air by combustion products. The acetylene is stored in a compressed gas cylinder as in a welding lab (although the gas is of a special purity for use in AA instruments). All precautions relating to compressed gas cylinders must be enforced; the cylinders must be secured to an immovable object, such as a wall, and they must have approved pressure regulators in place, etc. Tubing and connectors must be free of gas leaks. There must be an independently vented fume hood in place over the flame to take care of toxic combustion products. Volatile flammable organic solvents and their vapors, such as ether and methylene chloride, must not be present in the lab when the flame is lit.

Precautions should be taken to avoid flashbacks. Flashbacks result from improperly mixed fuel and air, such as when the flow regulators on the instrument are improperly set or when air is drawn back through the drain line of the premix burner (see Section 7.2). Manuals supplied with the instruments when they are purchased give more detailed information on the subject of safety.

Finally, periodic cleaning of the burner head and nebulizer is needed to ensure minimal noise level due to impurities in the flame. Scraping the slot in the burner head with a sharp knife or razor blade to remove carbon deposits and/or removing the burner head for the purpose of cleaning in an ultrasonic bath are two commonplace maintenance chores. The nebulizer should be dismantled, inspected, and cleaned periodically to remove impurities that may be collected there.

7.5.3c Sensitivities, Detection Limits, and Analytical Uses

Sensitivity and detection limits have specific definitions in AA. Sensitivity is defined as the concentration of an element which will produce an absorption of 1%. It is the smallest concentration that can be determined with a reasonable degree of accuracy. Detection limit is the concentration which gives a readout level that is double the noise level inherit in the experiment. It is a qualitative parameter in the sense that it is the minimum concentration that can be detected, but not accurately determined, like a "blip" on a noisy baseline. It would tell the analyst that the element is present, but not necessarily at an accurately determinable concentration level. Some sensitivities and detection limits for selected elements are given in Table 7.3.

The analytical uses of flame AA are very numerous. As indicated in

Table 7.3   Some Sensitivities and Detection Limits of Selected Elements in Flame AA at the Primary Wavelength Using an Air/Acetylene Flame and Both a Flow Spoiler and a Glass Impact Bead in the Premix Chamber

Element	Sensitivity (ppm)	Detection Limit (ppm)
Silver	0.03	0.0009
Arsenic	0.51	0.14
Calcium	0.08	0.001
Cadmium	0.02	0.0005
Chromium	0.04	0.002
Copper	0.03	0.001
Iron	0.04	0.003
Magnesium	0.003	0.0001
Manganese	0.03	0.001
Mercury	2.2	0.17
Nickel	0.04	0.004
Lead	0.19	0.01
Zinc	0.011	0.0008

Courtesy of Perkin-Elmer Corporation, Norwalk, CT.

Section 7.1, the vast majority of applications of atomic techniques is to solutions of metal ions. Thus, whenever metals are to be determined in a sample, atomic techniques are likely to be the chosen. Flame AA is the most widely used of these at the present time, probably because it has been around a long time and the instruments are less expensive than those which perhaps offer important advantages (see later sections). Table 7.4 lists some of the more common analytical uses of flame AA as well as other AA techniques.

7.6 GRAPHITE FURNACE AND OTHER ATOMIZERS

There are some limitations to the use of flames as atomizers compared to other atomizers that have been invented. First, they require relatively large sample volumes (at least several milliliters). Second, they produce fewer atoms in the path of the light, making them less sensitive. Third, they are not applicable to samples in certain matrices, due to undesirable matrix effects. This section discusses the alternative methods and especially emphasizes the graphite furnace.

Table 7.4   Examples of Analytical Uses for Atomic Absorption

Field of Endeavor	Examples of Samples and Analytes
Agriculture	Calcium, copper, iron, magnesium, manganese, potassium, sodium and zinc in soils, plant tissue, feeds and fertilizers
Biochemistry	Sodium, potassium, iron, lithium, copper, zinc, gold, lead, calcium, and magnesium in biological fluids, tissue, fingernails and hair
Environment	Calcium, copper, lithium, magnesium, manganese, potassium, sodium, strontium, and zinc in seawater and natural water, and other metals in airborne particulates
Food	Various metals in food and food additives, meat, seafood, cooking oils, and beverages
Heavy industry	Various metals in cement, coal ash, glass, ceramics, paint and paint additives, etc.
Metallurgy	Various minor elements in alloys, such as steels, brasses, alloys of aluminum, magnesium, lead, tin, copper, titanium, nickel, and iron, as well as plating solutions
Petroleum	Various metals in lubricating oils and additives, various wear metals in used lubricating oils, lead in gasoline, and metals in other fuels, oils and additives
Pharmaceuticals	Various metals in pharmaceutical preparations

Compiled from "Analytical Methods for Atomic Absorption Spectrophotometry", Perkin-Elmer Corporation, Norwalk, CT.

7.6.1 Graphite Furnace Atomizer

The graphite furnace method of atomization is useful and well established. The conversion of ions to atoms occurs in a small hollow graphite cylinder (tube) which replaces the burner head in flame AA. (The flame and graphite furnace are interchangeable in most instruments.) This cylinder is positioned horizontally such that the light beam passes directly through the center of it. There is a small hole in the center of the top side of the cylinder for introducing the sample. The sample is not introduced through any form of aspiration, but rather by means of an injection through this small hole, often onto a platform inside the tube. Thus, atoms are in the path of the light for a limited time defined by the sample volume. Very small and carefully measured sample volumes, 5–

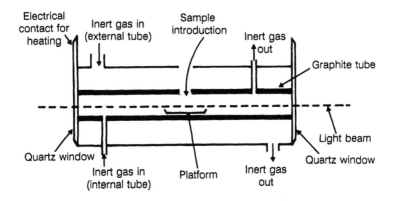

FIGURE 7.13   The graphite furnace atomizer.

50 µL, are used. This small size can be a distinct advantage when only small volumes are available. The tube is encased in a larger tube in order to facilitate protection against air oxidation of the graphite at the high temperature needed for atomization. An inert gas (argon or nitrogen) is fed both into the larger tube (but outside the inner tube) and into the inner tube. To further protect the tube and to prevent the sample from permeating through, a coating of pyrolytic graphite is often applied. Contact rings and quartz windows are attached at both ends of the tubes so as to seal the entire system (see Figure 7.13).

The tube itself acts as a heating element. The contact rings referred to above serve to make electrical contact to a power source. The power is controlled according to a three-step temperature program such that the sample is first dried, then ashed, and finally atomized at a temperature of about 2500 K. The atomic vapor fills the tube at this temperature, and light from the hollow cathode lamp is absorbed with the absorbance measured as in flame AA. The drying and ashing steps are required to destroy organic matter which produces smoke inside and would otherwise scatter the light from the source unless it is first flushed from the system by the flowing inert gas. The programming of the temperature then produces the time required for this to take place prior to the atomization.

Since there is a limited amount of sample present and since the atomic vapor is also eventually swept from the tube by the flowing inert gas, the absorbance signal is transient. This means that the recording of the absorbance is usually done as a function of time, such as on a recorder or by computer data acquisition, and the transient signal then appears as a sharp peak at a particular time.

The advantages of the graphite furnace are (1) the fact that only small sample volumes are needed, as previously stated, and (2) the sensitivity is substantially increased. Atomization in this furnace is nearly 100% efficient which is a tremendous improvement over the flame (0.1%). Detection limits are improved by as much as 1000 times. The major disadvantages are (1) there often can be matrix and background absorption effects, and (2) reproducibility is not as good as with flames. Standard additions (Chapter 5) can be a solution to the matrix effect. Continuum source correction and Zeeman Effect correction (utilizing a magnetic field) have been used effectively to reduce background effects. Detailed discussions of these are beyond the scope of this book.

7.6.2 Vapor Generation Methods

An alternative room temperature atomization method has been devised for mercury known as the cold vapor mercury technique. Also, an alternative technique for the difficult elements arsenic, bismuth, germanium, lead, antimony, selenium, tin, and tellurium called the hydride generation technique has been devised. These are collectively referred to as vapor generation techniques. They utilize a chemical reduction (of the metal ion to the metal atom or to the metal hydride) process rather than a direct atomization in a flame or furnace.

In the cold vapor technique for mercury, mercury atoms are generated via chemical reduction with a strong reducing agent, such stannous chloride. A stream of nitrogen or air sweeps the mercury atoms into a sample absorption cell that is positioned in the path of the light. A transient absorption signal, similar to the graphite furnace signal, is observed and measured.

In the hydride generation technique, sodium borohydride is the usual reducing agent. Again the reaction product is swept into the path of the light, but, unlike the mercury technique, the analyte is not yet in the free atomic state at this stage, but rather in the molecular hydride state. Once in the sample absorption cell, this vapor is heated with either a cool air/acetylene flame or with an electrical heating unit to create the atoms.

Advantages of vapor generation for the above listed elements can include (1) a decrease in the effects of sample matrices, (2) improved precision, and (3) higher sensitivity.

7.6.3 The Delves Cup

Improvements for some elements can be achieved by simply holding the sample in a cup in the flame (no aspiration). This cup, called the "Delves Cup," is composed of nickel. The atomic vapor generated is channeled into a sample absorption cell and the absorbance measured. Such a technique can utilize a sample as small as 100 µL (an advantage), but often suffers from poorer precision.

7.7 INDUCTIVELY COUPLED PLASMA

A technique that is gaining in popularity due to some important advantages over the techniques mentioned so far in this chapter is the Inductively Coupled Plasma technique better known as ICP. ICP is strictly an atomic emission technique most closely related to the flame photometry technique described earlier. The ICP emission source is not a flame, however, but a plasma. A plasma source is a flame-like system of ionized, very hot flowing argon gas that is directed through a quartz tube wrapped with a copper wire (coil). Radio frequency energy is applied to the coil creating an intense magnetic field inside the tube. The ionization of the gas is initiated by a "Tesla" coil, or spark, upstream from the quartz tube, causing the argon to become conductive, creating argon ions and electrons in the flow path. As the partially ionized argon flows through the quartz tube and the RF generated magnetic field, more ionization occurs and the gas becomes extremely hot — 9000–10,000 K. What emerges from the quartz tube then is a very hot spray of ionized argon that is the ICP source, giving the appearance of a flame.

The sample and standards, in the form of solutions (as with the other atomic techniques described in this chapter), are introduced into the flowing argon by aspiration just prior to the initial ionization. The entire system is pictured in Figure 7.14.

As stated previously in this chapter, a hotter source increases both atomization efficiency and excitation efficiency. Thus the hotter an excitation source, the more intense one would expect the emission to be, and the smaller the concentrations of metal ions that can be detected and accurately measured. ICP is therefore much more sensitive than all other atomic techniques. In addition, the concentration range over which the

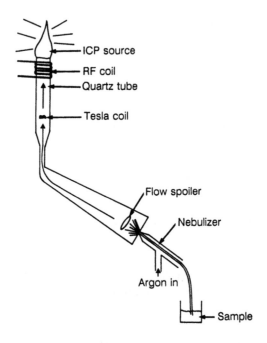

FIGURE 7.14   The ICP system described in the text.

emission intensity is linear is broader and simultaneous "multielement" analysis of a sample is possible. The instruments are more costly than AA and other systems.

7.8 OTHER ATOMIC EMISSION TECHNIQUES

7.8.1 Arc or Spark Emission Spectrography

A technique which utilizes a solid sample for light emission is Arc of Spark Emission Spectrography. In this technique, a high voltage is used to excite a solid sample held in an electrode cup in such a way that when a spark or arc jumps from this electrode to another electrode in this arrangement, atomization, excitation, and emission occur and the emitted light again is measured. The usual configuration is such that the emitted light is dispersed and detected with the use of photographic film, hence the name "spectrography." The "picture" that results is that of a combined

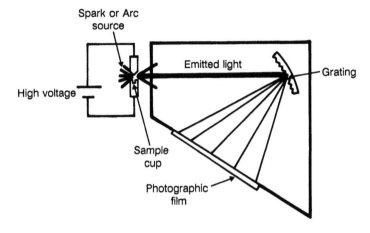

FIGURE 7.15   A diagram of a spark emission instrument.

line spectrum of all the elements in the sample. Identification (qualitative analysis) is then possible by comparing the locations of the lines on the film to location of lines on a standard film. Figure 7.15 shows the instrumental arrangement.

7.8.2 Atomic Fluorescence

Finally, we briefly describe a technique based on both absorption and emission — atomic fluorescence. When atoms that have been elevated to higher energy levels return to the ground state, the pathway could take them to some intermediate electronic states prior to the final drop. Such a series of drops back to the ground state, if accompanied by light emission, is a form of fluorescence, in this case, atomic fluorescence. (See Chapter 6 for a discussion of molecular fluorescence.) As with molecular fluorescence, the intensity of this emitted light is measured at right angles to the incident light and related to concentration.

7.9 SUMMARY OF ATOMIC TECHNIQUES

Table 7.5 summarizes the description and application of the techniques discussed in this chapter.

Table 7.5 A Summary of the Techniques Described in This Chapter

Technique	Principle	Comments
Flame photometry	Intensity of light emitted by atoms in a flame is measured and related to concentration.	A well-established technique that is used for only a limited number of analytes and samples due to the applicability and sensitivity of other techniques.
Flame AA	Light absorbed by atoms in a flame is measured and related to concentration.	A well-established technique that remains very popular for a large number of samples and analytes. Instruments are inexpensive.
Graphite furnace	Transient light absorbance signal from sample atomized in a graphite furnace (no flame) is measured and related to concentration.	A newer technique that has become popular due to its greater sensitivity and applicability to smaller sample size. Suffers from lack of precision.
Vapor generation	Light absorbed by chemically generated atomic vapor is measured and related to concentration.	Excellent technique for a limited number of analytes that are difficult to measure otherwise.
ICP	Light emitted by atoms in an inductively coupled plasma is measured and related to concentration.	A newer technique that has become popular due to advantages in sensitivity, linear range, and multielement analysis. Instruments are costly.
Spark or Arc emission	Light emitted by powdered solid sample caught in an arc or spark between a pair of electrodes is measured for qualitative analysis.	An excellent technique for qualitative analysis of metals in solids.
Atomic fluorescence	Light emitted by atoms in a flame excited by the absorption of light is measured and related to concentration.	Not a popular technique, but it can offer sensitivity advantages.

CHAPTER 8

ANALYTICAL SEPARATIONS

8.1 INTRODUCTION

Modern day chemical analysis can involve very complicated material samples — complicated in the sense that there can be many, many substances present in the sample creating a myriad of problems with interferences when the lab worker attempts the analysis. These interferences can manifest themselves in a number of ways. The kind of interference that is most familiar is probably one in which substances other than that the analyte generate an instrumental readout similar to the analyte, such that the interference adds to the readout of the analyte, creating an error. However, an interference can also suppress the readout for the analyte (e.g., by reacting with the analyte). An interference present in a chemical to be used as a standard (such as a primary standard) would cause an error, unless its presence and concentration were known. Analytical chemists must deal with these problems, and chemical procedures designed to effect separations or purification are now commonplace.

This chapter, and also Chapters 9 and 10, describe modern analytical separation science. First, purification procedures known as recrystallization and distillation will be described. Then, the separation techniques of extraction and chromatography are discussed. This is followed by, in Chapters 9 and 10, instrumental chromatography techniques which can resolve very complicated samples and quantitate usually in one easy step.

8.2 RECRYSTALLIZATION

Recrystallization is a purification technique for a solid, usually organic. The separation is based on the solid's solubility in a liquid solvent. A solvent is chosen such the solid is sparingly soluble at room temperature, but whose solubility increases considerably at higher temperatures. Both soluble and insoluble impurities are considered to be present, and the procedure removes both if their concentrations are not too large.

The key to the procedure is to use a *minimum* amount of solvent, such that the solid will just dissolve at the elevated temperature (usually the boiling point, if the solid is stable at that temperature). While maintaining this elevated temperature, any impurity that has not dissolved can be filtered out. The "insoluble" impurities are thus removed.

Soluble impurities, however, are still present in the filtrate. This is where the minimum amount of solvent comes into play. The procedure calls for the temperature to be lowered back to the original value. The soluble impurities will stay dissolved if their solubility has not been exceeded (if they are present in a small amount). However, the solid being purified will have its solubility exceeded. Since the minimum amount of solvent was used to just dissolve the solid at the elevated temperature, it will thus precipitate from the solution. The solid, presumably purified, can then be filtered and dried. It may be necessary to perform the recrystallization several times in order to get the desired purity.

8.3 DISTILLATION

Distillation is a method of purification of liquids contaminated with either dissolved solids or miscible liquids. The method consists of boiling and evaporating the mixture followed by recondensation of the vapors in a "condenser," which is a tube cooled by isolated, cold tap water. The theory is that the vapors (and thus recondensed liquids) will be purer than the original liquid. The separation is based on the fact that the contaminants have different boiling points and vapor pressures than the liquid to be purified. Thus, when the liquid is boiled and evaporated, the vapors (and recondensed liquids) created have a composition different from the original liquid. The substances with lower boiling points and higher vapor pressures are therefore separated from substances that have high boiling points and low vapor pressures.

Distillation of water to remove hardness minerals is an example and probably the most common application in an analytical laboratory. Of all the applications of distillation, it is one of the easiest to perform. While water is known to have a relatively high boiling point and low vapor pressure, the dissolved minerals are ionic solids that generally have extremely high boiling points (indeed, extremely high melting points) and extremely low vapor pressures. Thus, a simple distilling apparatus and a single distillation, or, at most, two (doubly distilled) or three (triply distilled) distillations, will produce very pure water.*

Organic liquids that are contaminated with other organic liquids usually constitute a much more difficult situation. Such liquids probably have such similar boiling points and vapor pressures that a distillation of a mixture of two or more would result in all being present in the distillate (the condensed vapors) — an unsuccessful purification. However, the liquid that has the highest vapor pressure and/or lowest boiling point, while not being completely purified, would be present in the distillate at a higher concentration level than the other components. It follows that if the distillate were then to be redistilled, perhaps over and over again, further enrichment of this component would take place such that an acceptable purity would eventually be obtained. However, the time involved in such a procedure would be prohibitive. A procedure known as "fractional distillation," solves the problem.

Fractional distillation involves repeated evaporation/condensation steps before the distillate is actually collected. These repeated steps occur in a "fractionating column" (tube) above the original heated container — a column that contains a high surface area inert material for condensing the vapors. As the vapors condense on this material, the material itself heats up and the condensate reevaporates. The reevaporated liquid then moves further up the column, contacts more cold inert material, and the

---

*   Water is often "deionized" using an ion exchange (Section 8.6) cartridge to remove hardness minerals. Such deionization is often done in conjuction with distillation, such that the water is both deionized and distilled prior to use. Also, if the water is contaminated with organics, or other low boiling substances, a charcoal filter cartridge is often used as well.

process occurs again and again and again as the liquid makes its way up the column. If a fractionating column were used that is long enough and contains a sufficient quantity of the high surface area material, any purification based on differences in boiling point and vapor pressure can be affected. A schematic diagram of a distillation apparatus fitted with a fractionating column is shown in Figure 8.1. The high surface area packing material in a fractionating column typically consists of glass beads, glass helices, or glass wool.

Each time a single evaporation/condensation step occurs in a fractionating column, the condensate has passed through what has been called a "theoretical plate." A theoretical plate is thus that segment of a fractionating column in which one evaporation/condensation step occurs. The name is derived from the concepts in which the condensate is actually captured on small "plates" inside the fractionating column from which it is again boiled and evaporated. A fractionating column used for a given liquid mixture is then identified as having a certain number of theoretical plates, and given liquid mixtures are known to require a certain number of theoretical plates in order to achieve a given purity. The "height equivalent to a theoretical plate," or HETP, is the length of fractionating column corresponding to one theoretical plate. If the number of theoretical plates required is known, then the analyst can select a height of column that would contain the proper number of plates according to manufacturer's specifications or according to his own measurements of a homemade column. Height selection is not entirely experimental, however. The use of liquid-vapor composition diagrams to predict the theoretical plates required can help. These diagrams are based on boiling point and vapor pressure differences in a pair of liquids. Further discussion of the use of these diagrams is beyond the scope of this book.

8.4 LIQUID-LIQUID EXTRACTION

8.4.1 Introduction

One popular method of separating an analyte species from a complicated liquid sample is the technique known as "liquid-liquid extraction" or "solvent extraction." In this method, the sample containing the analyte is a liquid solution, typically a water solution, that also contains other

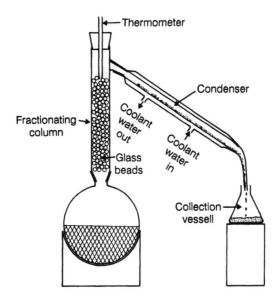

FIGURE 8.1   A schematic diagram of a distillation assembly complete with a fractionating column, condenser, and collection vessel.

solutes. The need for the separation usually arises from the fact that the other solutes, or perhaps the original solvent, interfere in some way with the analysis technique chosen. An example is a water sample that is being analyzed for a pesticide residue. The water may not be a desirable solvent, and there may be other solutes that may interfere. It is a "selective dissolution" method; a method in which the analyte is removed from the original solvent and subsequently dissolved in a different solvent, (extracted) while most of the remainder of the sample remains unextracted, i.e., remains behind in the original solution.

The technique obviously involves two liquid phases — one the original solution and the other the extracting solvent. The important criteria for a successful separation of the analyte are (1) that these two liquids be immiscible and (2) that the analyte be more soluble in the extracting solvent than the original solvent.

8.4.2 The Separatory Funnel

The extraction takes place in a specialized piece of glassware known as a separatory funnel. The separatory funnel is manufactured especially for solvent extraction. It has a "teardrop" shape with a stopper at the top and a stopcock at the bottom (Figure 8.2). The sample and solvent are placed together in the funnel, the funnel is tightly stoppered and, while holding the stopper in with the index finger, shaken vigorously for a moment. Following this, the funnel may need to be vented, since one of the liquids is likely to be a volatile organic solvent, such as methylene chloride. Venting is accomplished by opening the stopcock when inverted. This shaking/venting step is then repeated several times such that the two liquids have plenty of opportunity for the intimate contact required for the analyte to pass into the extracting solvent to the maximum possible extent. See Figure 8.3 for shaking/venting illustrations. Following this procedure, the funnel is positioned in a padded ring in a ring stand (Figure 8.2) and left undisturbed for a moment to allow the two immiscible layers to once again separate. The purpose of the specific design of the separatory funnel is mostly to provide for easy separation of the two immiscible liquid layers after the extraction takes place. All one needs to do is remove the stopper, open the stopcock, allow the bottom layer to drain, then close the stopcock when the interface between the two layers disappears from sight in the stopcock. The denser of the two liquids is the bottom layer and will be drained through the stopcock first. The entire process may need to be repeated several times, since the extraction is likely not to be quantitative. This means that another quantity of fresh extracting solvent may need to be introduced into the separatory funnel with the sample and the shaking procedure repeated. Even so, the experiment may never be completely quantitative. See the next section for the theory of extraction and a more in-depth discussion of this problem.

8.4.3 Theory

The process of a solute dissolved in one solvent being "pulled out," or "extracted," into a new solvent actually involves an equilibrium process. At the time of initial contact, the solute will move from the original solvent to the extracting solvent at a particular rate, but, after a time, it will begin to move back to the original solvent at a particular rate. When the two rates are equal, we have equilibrium. We can thus write the following:

$$A_{orig} \rightleftharpoons A_{ext} \qquad (8.1)$$

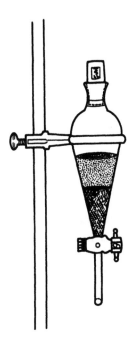

FIGURE 8.2   A separatory funnel containing two immiscible liquids.

FIGURE 8.3   Illustrations of the positioning of the hands and the procedure for the shaking and venting mentioned in the text.

in which "A" refers to "analyte," and "orig" and "ext" refer to "original solvent" and "extracting solvent," respectively. If the analyte is more soluble in the extracting solvent than in the original solvent, then, at equilibrium, a greater percentage will be found in the extracting solvent and less in the original solvent. If the analyte is more soluble in the original solvent, then the greater percentage of analyte will be found in the original solvent. Thus, the amount that gets extracted depends on the relative distribution between the two layers, which, in turn, depends on the solubilities in the two layers. A distribution coefficient analogous to an equilibrium constant (also called the "partition coefficient") can be defined as follows:

$$K = \frac{[A]_{ext}}{[A]_{orig}} \qquad (8.2)$$

Often, the value of K is approximately equal to the ratio of the solubilities of "A" in the two solvents. If the value of K is very large, the transfer of solute to the extracting solvent can be considered to be quantitative. A value around 1.0 would indicate equal distribution, and a small value would indicate very little transfer.

Uses of the distribution coefficient include (1) the calculation of the amount of a solute that is extracted in a single extraction step, (2) the determination of the weight of the solute in the original solute (important if you are quantitating the solute in this solvent), (3) the calculation of the optimum volume of both the extracting solvent and the original solution to be used, (4) the number of extractions needed to obtain particular quantity or concentration in the extracting solvent, and (5) the percent extracted. The following expansion of Equation 8.2 is useful for these.

$$K = \frac{\dfrac{W_{ext}}{V_{ext}}}{\dfrac{W_{orig}}{V_{orig}}} \qquad (8.3)$$

in which W_{ext} is the weight of the solute extracted into the extracting solvent, V_{ext} is the volume of the extracting solvent used, W_{orig} is the weight of the solute in the original solvent, and V_{orig} is the volume of the original solvent used.

More complicated chemical systems may require a more universally applicable quantity called the "distribution ratio" to describe the system. These involve situations in which the analyte species may be found in different chemical states and different equilibrium species, some of which may be extracted while others are not extracted. An example is an equi-

equilibrium system involving a weak acid. In such a system, there may be one (or several) protonated species and one unprotonated species. The "distribution ratio," D, then takes into account all analyte species present.

$$D = \frac{C_A(\text{ext})}{C_A(\text{orig})} \tag{8.4}$$

in which C_A (ext) and C_A (orig) represent the total concentration of all analyte species present in the two phases regardless of chemical state. Further treatment of this situation is beyond the scope of this text.

8.4.4 Countercurrent Distribution

Just one extraction performed on a solution of a complicated sample will likely not result in total or at least sufficient separation of the analyte from other interfering solutes. Not only will these other species also be extracted to a certain degree with the analyte, but some of the analyte species will likely be left behind in the original solvent as well. Thus, the analyst will need to perform additional extractions on both the extracting solvent, to remove the other solutes that were extracted with the analyte, and the original solution, to remove additional analyte that was not extracted the first time. One can see that dozens of such extractions may be required to achieve the desired separation. Eventually, there would be a separation, however. The process is called countercurrent distribution.

In countercurrent distribution, the extracting solvent, after first being in contact with the original solution, is moved to another separatory funnel in which there is fresh original solvent, while fresh extracting solvent is brought into the original funnel and the extractions performed. Then, the extracting solvent from the second funnel is moved to a third funnel containing fresh original solvent, the extracting solvent from the first is moved to the second funnel, and fresh extracting solvent is introduced into the first funnel. The process continues in this manner until the desired separation occurs. The concept is illustrated in Figure 8.4. The top half in each segment (a, b, etc.) represents extracting solvent, while the lower half represents original solvent. A mixture of "x" and "." is being separated. In each segment, fresh extracting solvent is introduced

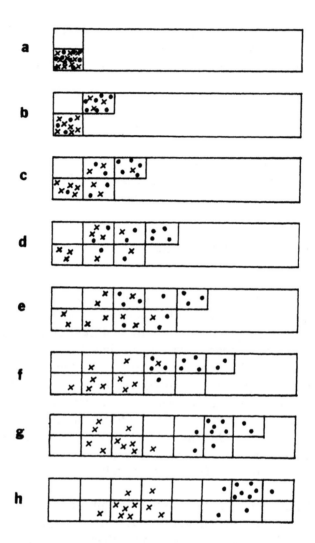

FIGURE 8.4   A diagram depicting countercurrent distribution as discussed in the text. (From Kenkel, J., *Analytical Chemistry for Technicians*, Lewis Publishers, Inc., Chelsea, MI, 1988. With permission.)

on the upper left, while fresh original solvent is introduced on the lowered right. This illustration shows a complete separation in segments "g" and "h." Some chromatography methods are based on this concept and will be discussed in Section 8.5.

8.4.5 Concentrators

Following an extraction procedure, it is often necessary to evaporate a quantity of the extracting solvent in order to increase the concentration of the analyte prior to gas chromatographic analysis (Chapter 9), for example. Two so-called "concentrators" in common use are the rotary evaporator and the Kuderna-Danish evaporative concentrator. With the rotary evaporator, the analyte/solvent mixture is heated in a round bottom flask which is rotated while lying nearly on its side in a steam bath. Solvent vapors are condensed and collected in a separate flask under a condenser off to the side, while the analyte remains in the rotating flask. The result is a solution of the analyte that has a higher concentration. The procedure takes place while the system is under reduced pressure. Bumping and superheating are avoided because of the rotation.

The Kuderna-Danish apparatus is a distillation/reflux assembly that utilizes a special fractionating column called the Snyder column. The lighter lower-boiling solvent(s) escapes the Snyder column to a solvent recovery condenser, while the analyte refluxes back to the original container, the Kuderna-Danish flask, and graduated concentrator tube. The end result is a 5–10 mL fraction of analyte concentrated in this tube.

Both of these units are used when the analyte concentration is too low to be detected otherwise.

8.5 LIQUID-SOLID EXTRACTION

There are instances in which the analyte will need to be extracted from a solid material sample rather than a liquid. As in the above discussion for liquid samples, such an experiment is performed either because it is not possible or necessary to dissolve the entire sample or because it is undesirable to do so because of interferences that may also be present. In these cases, the weighed solid sample, preferably finely divided, is brought into contact with the extracting liquid in an appropriate container (not a separatory funnel) and usually shaken or stirred for a period of time such that the analyte species is removed from the sample and dissolved in the liquid. The time required for this shaking is determined by the rate of the dissolving. A separatory funnel is not used since two liquid phases are not present, but rather a liquid and a solid phase. A simple beaker, flask, or test tube usually suffices.

Following the extraction, the undissolved solid material is then filtered out and the filtrate analyzed. Examples of this would be soil samples to be analyzed for metals, such as potassium or iron, and cellophane or insulation samples to be analyzed for formaldehyde residue. The extracting liquid may or may not be aqueous. Soil samples being analyzed for metals, for example, utilize aqueous solutions of appropriate inorganic compounds, sometimes acids, while soil samples, or the cellophane or insulation samples referred to above, that are being analyzed for organic compounds utilize organic solvents for the extraction. As with the liquid-liquid examples, the extract is then analyzed by whatever analytical technique is appropriate — atomic absorption for metals and spectrophotometry or gas or liquid chromatography for organics. Concentration methods, as with the Kuderna-Danish apparatus described in Section 8.4, may also be required, especially for organics.

It may be desirable to try to keep the sample exposed to *fresh* extracting solvent as much as possible during the extraction in order to maximize the transfer to the liquid phase. This may be accomplished by pouring off the filtrate and reintroducing fresh solvent periodically during the extraction and then combining the solvent extracts at the end. There is a special technique and apparatus, however, that has been developed, called the Soxhlet extraction apparatus, which accomplishes this automatically. The Soxhlet apparatus is shown in Figure 8.5. The extracting solvent is placed in the flask at the bottom, while the weighed solid sample is place in the solvent-permeable thimble in the compartment directly above the flask. A condenser is situated directly above the thimble. The thimble compartment is a sort of cup that fills with solvent when the solvent in the flask is boiled, evaporated, and condensed on the condenser. The sample is thus exposed to freshly distilled solvent as the cup fills. When the cup is full, the glass tube next to the cup is also full, and when it (the tube) begins to overflow, the entire contents of the cup is siphoned back to the lower chamber and the process repeated. The advantages of such an apparatus are (1) fresh solvent is continuously in contact with the sample (without having to introduce more solvent, which would dilute the extract) and (2) the experiment takes place unattended and can conveniently occur overnight if desired.

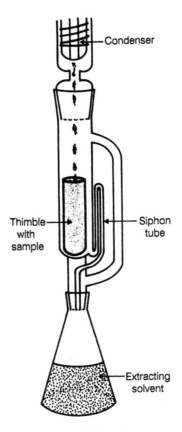

FIGURE 8.5   A drawing of a Soxhlet Extraction apparatus.

8.6 CHROMATOGRAPHY

8.6.1 Introduction

A myriad of techniques used to separate complex samples comes under the general heading of "chromatography." The nature of chromatography allows much more versatility, speed, and applicability than any of the other techniques discussed in this chapter, particularly when the modern instrumental techniques of gas chromatography (GC) and high performance liquid chromatography (HPLC) are considered. These latter techniques are covered in detail in Chapters 9 and 10. In this chapter, we introduce the general concepts of chromatography and give a perspective on its scope. Since there are many different classifications, this will include an organizational scheme covering the different types and configurations that exist.

Chromatography is the separation of the components of a mixture based on the different degrees to which they interact with two separate material phases. The nature of the two phases and the kind of interaction can be varied, and this gives rise to the different "types" of chromatography which will be described in the next section. One of the two phases is a moving phase (the "mobile" phase) while the other does not move (the "stationary" phase). The mixture to be separated is usually introduced into the mobile phase which then is made to move or percolate through the stationary phase either by gravity or some other force. The components of the mixture are attracted to and slowed by the stationary phase to varying degrees, and as a result, they move along with the mobile phase at varying rates and are thus separated. Figure 8.6 illustrates this concept.

The mobile phase can be either a gas or a liquid, while the stationary phase can be either a liquid or solid. One classification scheme is based on the nature of the two phases. All techniques which utilize a gas for the mobile phase come under the heading of "gas chromatography" (GC). All techniques that utilize a liquid mobile phase come under the heading of "liquid chromatography" (LC). Additionally, we have gas-liquid chromatography (GLC), gas-solid chromatography (GSC), liquid-liquid chromatography (LLC), and liquid-solid chromatography (LSC) if we wish to stipulate the nature of the stationary phase as well as the mobile phase. It is more useful, however, to classify the techniques according to the nature of the interaction of the mixture components with the two phases. These classifications we refer to as "types" of chromatography.

8.6.2 "Types" of Chromatography

8.6.2a Partition

In Section 8.4, we stated that some chromatography methods are based on the concept of countercurrent solvent extraction. You will recall that this is the technique in which a large number of extractions are performed, with fresh extracting solvent being brought into contact with

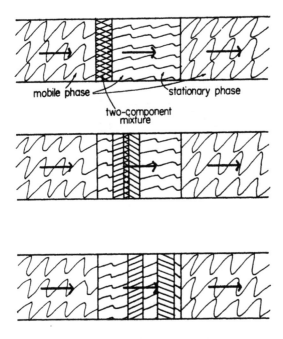

FIGURE 8.6   Mixture components separate as they move through the stationary phase with the mobile phase. (From Kenkel, J., *Analytical Chemistry for Technicians*, Lewis Publishers, Inc., Chelsea, MI, 1988. With permission.)

previously extracted samples, while fresh sample solvent is brought into contact with solvent extract from previous extractions (see Figure 8.4 and accompanying discussion). The extracting solvent can be thought of as continuously "moving" across the sample solvent, while the latter remains stationary. The mixture components, which were initially found in the first segment of sample solvent, then distribute back and forth between the two phases as the extracting liquid moves and are found individually separated at different points along the way according to their individual solubilities in the two solvents (see Figure 8.7).

The extracting solvent in this scenario is the chromatographic mobile phase, while the sample solvent is the stationary phase. Liquid-liquid partition chromatography (LLC) is based on this idea. The mobile phase is a liquid which moves through a liquid stationary phase as the mixture components "partition" or distribute themselves between the two phases and become separated. The separation mechanism is thus one of the dissolving of the mixture components to different degrees in the two phases according to their individual solubilities in each.

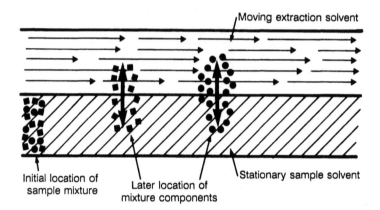

Moving extraction solvent

Initial location of sample mixture

Later location of mixture components

Stationary sample solvent

FIGURE 8.7   Illustration of the movement of an extracting solvent across a sample solvent and the separation of two compounds initially found in the sample solvent.

It may be difficult to imagine a liquid mobile phase used with a liquid stationary phase. What experimental setup allows one liquid to move through another liquid (immiscible in the first), and how can one expect partitioning of the mixture components to occur? The stationary phase actually consists of a thin liquid film either *adsorbed* or *chemically bonded* to the surface of finely divided solid particles as shown in Figure 8.8. Chemically bonded liquid stationary phases represent a fairly recent development in this area, as opposed to adsorbed liquid phases. The latter is considered to be a more "classical" chromatography technique. Bonded-phase chromatography (BPC) has a distinct stability advantage. It is not removed from the solid substrate either by reaction or by heat. BPC has become the most popular of the two by far.

Since the separation depends on the relative solubilities of the components in the two phases, the polarities of the components and that of the stationary and mobile phases are important to consider. If the stationary phase is somewhat polar, it will retain polar components more than it will nonpolar components, and thus the nonpolar components will move more quickly through the stationary phase than the polar components. The reverse would be true if the stationary phase were nonpolar. Of course, the polarity of a liquid mobile phase plays a role too.

The mobile phase for partition chromatography can also be a gas (GLC). In this case, however, the mixture components' solubility in the

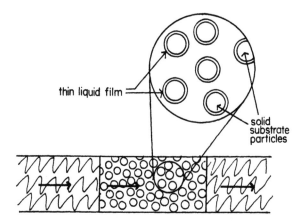

thin liquid film

solid
substrate
particles

FIGURE 8.8   Partition chromatography with a thin liquid film adsorbed or chemically bonded to the surface of finely divided solid particles. (From Kenkel, J., *Analytical Chemistry for Technicians*, Lewis Publishers, Inc., Chelsea, MI, 1988. With permission.)

mobile phase is not an issue — rather their relative vapor pressures are important. This idea will be expanded in Chapter 9.

In summary, partition chromatography is a type of chromatography in which the stationary phase is a liquid adsorbed or chemically bonded to the surface of a solid substrate, while the mobile phase is either a liquid or gas. The mixture components dissolve in and out of the mobile and stationary phases as the mobile phase moves through the stationary phase and the separation occurs as a result. Examples of mobile and stationary phases will be discussed in Chapters 9 and 10.

8.6.2b Adsorption

Another chromatography "type" is adsorption chromatography. As the name implies, the separation mechanism is one of adsorption. The stationary phase consists of finely divided solid particles packed inside a tube, but with no stationary liquid substance present to function as the stationary phase, as is the case with partition chromatography. Instead, the solid itself is the stationary phase and the mixture components, rather than dissolve in a liquid stationary phase, adsorb or "stick" to the surface of the solid. Different mixture components adsorb to different degrees

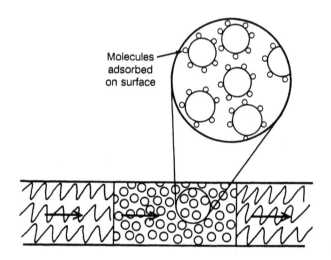

Molecules adsorbed on surface

FIGURE 8.9   A depiction of adsorption chromatography.

of strength, which also depends on the mobile phase, and thus again they become separated as the mobile phase moves. The nature of the adsorption involves the interaction of polar molecules or molecules with polar groups, with a very polar solid stationary phase. Thus, hydrogen bonding or similar molecule-molecule interactions are involved.

This "very polar solid stationary phase" is typically silica gel or alumina. The polar mixture components can be organic acids, alcohols, etc. The mobile phase can be either a liquid or a gas. This type of chromatography is depicted in Figure 8.9.

8.6.2c Ion-Exchange

A third chromatography "type" is ion-exchange chromatography. As the name implies, it is a method for separating mixtures of ions, both inorganic and organic. The stationary phase consists of very small polymer resin "beads" which have many ionic bonding sites on their surfaces. These sites selectively exchange ions with certain mobile phase compositions as the mobile phase moves. Ions that bond to the charged site on the resin beads are thus separated from ions that do not bond. Repeated changing of the mobile can create conditions which will further

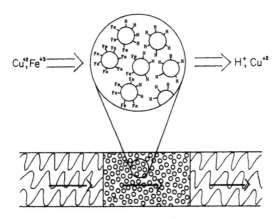

FIGURE 8.10   A depiction of ion-exchange chromatography. (From Kenkel, J., *Analytical Chemistry for Technicians*, Lewis Publishers, Inc., Chelsea, MI, 1988. With permission.)

selectively dislodge and exchange bound ions which then are also separated. This stationary phase material can be either an anion exchange resin, which possesses positively charged sites to exchange negative ions, or a cation exchange resin, which possesses negatively charge sites to exchange positive ions. The mobile phase can only be a liquid. Further discussion of this type can be found in Chapter 10. Figure 8.10 depicts ion-exchange chromatography.

8.6.2d Size-Exclusion

Size-exclusion chromatography, also called "gel-permeation" (GPC) or "gel-filtration" (GFC) chromatography, is a technique for separating dissolved species on the basis of their size. The stationary phase consists of porous polymer resin particles. The components to be separated can enter the pores of these particles and be slowed from progressing through this stationary phase as a result. Thus, the separation depends on the sizes of the pores relative to the sizes of the molecules to be separated. Small particles are slowed to a greater extent than larger particles, some of which may not enter the pores at all, and thus the separation occurs. The mobile phase for this type can also only be a liquid, and it too is discussed further in Chapter 10. The separation mechanism is depicted in Figure 8.11.

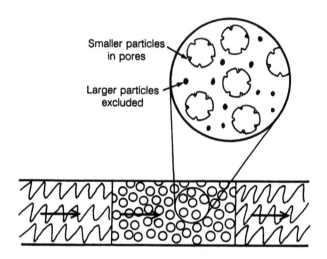

Smaller particles in pores

Larger particles excluded

FIGURE 8.11   A depiction of size-exclusion chromatography.

8.6.3 Chromatography Configurations

Chromatography techniques are further classified according to "configuration" — how the stationary phase is "contained"; how the mobile phase is configured with respect to the stationary phase in terms of physical state (gas or liquid); positioning; and how and in what direction the mobile phase travels in terms of gravity, capillary action, or other forces.

Configurations can be broadly classified into two categories: the "planar" methods and the "column" methods. The planar methods utilize a thin sheet of stationary phase material, and the mobile phase moves across this sheet, either upward ("ascending" chromatography), downward ("descending" chromatography) or horizontally ("radial" chromatography). Column methods utilize a cylindrical tube to contain the stationary phase, and the mobile phase moves through this tube either by gravity, with the use of a high pressure pump, or by gas pressure. Additionally, with the exception of paper chromatography, those that utilize a liquid for the mobile phase are capable of utilizing all of the "types" reviewed above. Paper chromatography utilizing unmodified cellulose sheets is strictly partition chromatography (see Section 8.6.3a). If the mobile phase is a gas (gas chromatography), the "type" is limited to adsorption and partition methods. Table 8.1 summarizes the different configurations. Let us consider each individually.

Table 8.1   **Chromatography Configurations and Their Applicable Options**

Geometry	Configuration	Migration Direction	Applicable Types
Planar	Paper	Ascending, descending, radial	Partition
Planar	Thin-layer	Ascending, descending, radial	Adsorption, partition, ion exchange, size exclusion
Column	Open-column	Descending	Adsorption, partition, ion exchange, size exclusion
Column	GC	N/A	Adsorption, partition
Column	HPLC	N/A	Adsorption, partition, ion-exchange, size-exclusion

8.6.3a Paper and Thin-Layer Chromatography

Paper chromatography and thin layer chromatography (TLC) constitute the planar methods mentioned above. Paper chromatography makes
use of a sheet of paper having the consistency of filter paper (cellulose)
for the stationary phase. Since such paper is hydrophilic, the stationary
phase is actually a thin film of water unintentionally adsorbed on the
surface of the paper. Thus, paper chromatography represents a form of
partition chromatography only. The mobile phase is always a liquid.

With thin-layer chromatography, the stationary phase is a thin layer
of material spread across a plastic sheet or glass or metal plate. Such plates
or sheets may either be purchased commercially already prepared or they
may be prepared in the laboratory. The thin-layer material can be any
of the stationary phases described earlier, and thus TLC can be any of
the four types, including adsorption, partition, ion-exchange, and size-
exclusion. Perhaps the most common stationary phase for TLC, however,
is silica gel, a highly polar stationary phase for adsorption chromatography,
as mentioned earlier. Also common is pure cellulose, the same material
for paper chromatography, and here also we would have partition
chromatography. The mobile phase for TLC is always a liquid.

The most common method of configuring a paper or thin-layer experiment is the ascending configuration shown in Figure 8.12. The mixture

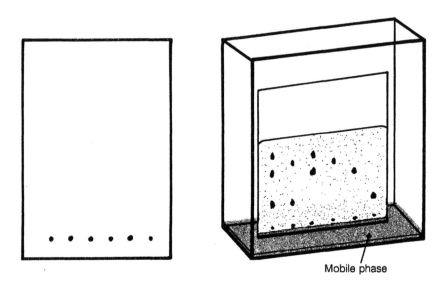

FIGURE 8.12   The paper or thin-layer chromatography configurations. The drawing on the left shows the paper or TLC plate with the spots applied. The drawing on the right shows the chromatogram in the developing chamber nearing complete development.

to be separated is first "spotted" (applied as a small "spot") within 1 in. of one edge of a 10 in. square rectangular paper sheet or TLC plate. A typical experiment may be an attempt to separate several spots representing different samples and standards on the same sheet or plate. Thus, as many as eight or more spots may be applied on one sheet or plate. So that all spots are aligned parallel to the bottom edge, a light pencil mark can be drawn prior to spotting. The size of the spots must be such that the mobile phase will carry the mixture components without streaking. This means that they must be rather small; they must be applied with a very small diameter capillary tube or micropipet. An injection syringe with a 25 μL maximum capacity is usually satisfactory.

Following spotting, the sheet or plate is placed spotted edge down in a "developing chamber" that has the liquid mobile phase in the bottom to a depth lower than the bottom edge of the spots. The spots must not contact the mobile phase. The mobile phase proceeds upward by capillary action and sweeps the spots along with it. At this point, chromatography is in progress; the mixture components will move with the mobile phase at different rates through the stationary phase and, if the mixture components are colored, evidence of the beginning of a separation is

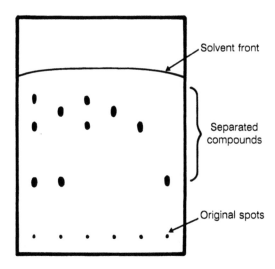

FIGURE 8.13   A developed paper of thin-layer chromatogram.

visible on the sheet or plate. The end result, if the separation is successful, is a series of spots along a path immediately above the original spot locations each representing one of the components of the mixture spotted there (see Figure 8.13).

If the mixture components are not colored, any of a number of techniques designed to make the spots visible may be employed. These include iodine staining in which iodine vapor is allowed to contact the plate. Iodine will absorb on most spots rendering them visible. Alternatively, a fluorescent substance may be added to the stationary phase prior to the separation (available with commercially prepared plates) such that the spots, viewed under an ultraviolet light, will be visible because they do not fluoresce, while the stationary phase surrounding the spots does fluoresce.

The visual examination of the chromatogram can reveal the identities of the components, especially if standards were spotted on the same paper or plate. Retardation factors (so called R_f factors) can also be calculated and used for qualitative analysis. These factors are based on the distance the mobile phase has traveled on the paper (measured from the original spot of the mixture) relative to the distances the components have traveled, each measured from either the center or leading edge of the original spot to the center or leading edge of the migrated spot.

$$R_f = \frac{\text{distance mixture component has traveled}}{\text{distance mobile phase has traveled}} \qquad (8.5)$$

These factors are compared to those of standards to reveal the identities of the components.

Quantitative analysis is also possible. The spot representing the component of interest can be cut (in the case of paper chromatography) or scraped from the surface (TLC) and dissolved and quantitated by some other technique, such as spectrophotometry. Alternatively, modern scanning densitometers, which utilize the measurement of the absorbance or reflectance of ultraviolet or visible light at the spot location, may be used to measure quantity.

Using the TLC concept to prepare pure substances for use in other experiments, such as standards preparation or synthesis experiments, is possible. This is called preparatory TLC and involves a thicker layer of stationary phase so that larger quantities of the mixture can be spotted and a larger quantity of pure component obtained.

Additional details of planar chromatography, such as methods of descending and radial development, how to prepare TLC plates, tips on how to apply the sample, what to do if the spots are not visible, and the details of preparatory TLC, etc., are beyond the scope of this text.

8.6.3b Classical Open-Column Chromatography

Another configuration for chromatography consists of a vertically positioned glass tube in which is placed the stationary phase. It is typical for this tube to be open at the top, to have an inner diameter on the order of 1 cm, and to have a stopcock at the bottom, making it similar to a buret in appearance. With this configuration, the mixture to be separated is placed at the top of the column and allowed to pass onto the stationary phase by opening the stopcock. The mobile phase is then added and continuously fed into the top of the column and flushed through by gravity flow. The mixture components separate on the stationary phase as they travel downward and, unlike the planar methods, are then collected as they elute from the column. In the classical experiment, a "fraction collector" is used to collect the eluting solution. A typical fraction collector consists of a rotating carousel of test tubes positioned under the column such that fractions of eluate are collected over a period of time,

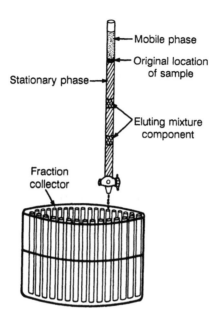

Mobile phase

Original location of sample

Stationary phase

Eluting mixture component

Fraction collector

FIGRUE 8.14   The classical open-column chromatography configuration with fraction collector.

such as overnight or a period of days, in individual test tubes (see Figure 8.14). This then makes qualitative or quantitative analysis possible through the analysis of these fractions by some other technique, such as spectrophotometry.

The length of the column is determined according to the degree to which the mixture components separate on the stationary phase chosen. Difficult separations would require more contact with the stationary phase and thus may require longer columns. Again, all four types (adsorption, partition, ion exchange, and size exclusion) can be used with this technique.

It is well known that classical open-column chromatography has been largely displaced by the modern instrumental techniques of liquid chromatography. However, open columns are still used where extensive sample "clean-up" in preparation for the instrumental method is necessary. One can imagine that "dirty" samples originating, for example, from animal feed extractions or soil extractions, etc. may have large concentrations of undesirable components present. Since only very small samples (on the order of 1–20 μL) are needed for the instrumental method, the time required for obtaining a clean sample by this method, assuming the components of interest are not retained, is the time it takes for an initial amount of mobile phase to pass through from top to bottom. Compared to the "overnight" time frame, such a clean-up time is quite minimal and does not diminish the speed of the instrumental methods.

8.6.3c Instrumental Chromatography

The concept of the finely divided stationary phase packed inside a column allowing the collection of the individual components as they elute as discussed in the last section presents a useful, more practical alternative. One can imagine such a column along with a continuous mobile phase flow system, a device for introducing the mixture to the flowing mobile phase, and an electronic detection system at the end of the column — all of this incorporated into a single unit (instrument) used for repeated, routine laboratory applications. There are two such chromatography configurations which are in common use today known as GC and HPLC. These techniques essentially can incorporate all types of column chromatography discussed thus far (HPLC), as well as those types in which the mobile phase can be a gas (GC). Both add a degree of efficiency and speed to the chromatography concept. HPLC, for example, is such a "high performance" technique for liquid mobile phase systems that a procedure that would normally take hours or days with open columns actually requires much less time. The full details of these instrumental techniques are discussed in Chapters 9 and 10.

8.7 ELECTROPHORESIS

Another separation technique is one which utilizes the varied rates and direction with which different dissolved ions migrate through the solution while under the influence of an electric field. This technique is called "electrophoresis." "Zone Electrophoresis" refers to the common case in which a medium such as cellulose or gel is used to contain the solution. A schematic diagram of the electrophoresis apparatus resembles an electrochemical apparatus in many respects. A power supply is needed for connection to a pair of electrodes to create the electric field. The medium and sample to be separated are positioned between the electrodes.

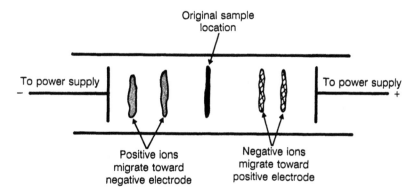

FIGURE 8.15   Illustration of the concept of electrophoresis.

The basic concept of the technique and apparatus is illustrated in Figure 8.15.

Electrophoresis is for separating ions, since only ions will migrate under the influence of an electric field; negative ions migrate to the positive electrode and positive ions to the negative electrode. Scientists have found electrophoresis especially useful in biochemistry experiments in which charged amino acid molecules and other biomolecules need to be separated. Thus, application to protein and nucleic acid analysis has been popular.

The principles of separation are (1) ions of opposite charge will migrate in different directions and become separated on that basis and (2) ions of like charge, while migrating in the same direction, become separated due to different migration rates. Factors influencing migration rate are charge values (i.e., −1 as opposed to −2, for example) and/or different mobilities. The mobility of an ion is dependent on the size and shape of the ion, as well as the nature of the medium through which it must migrate. The biomolecules referred to above can vary considerably in size and shape, and thus electrophoresis is a powerful technique for separating them. As for the medium used, there are some options, including the use of an electrolyte-soaked cellulose sheet (paper electrophoresis) or a thin gel slab (gel electrophoresis), as well as the nature of the electrolyte solution used and its pH.

Figure 8.16 represents a paper electrophoresis apparatus. The soaked cellulose sheet is sandwiched between two horizontal glass plates with the ends dipped into vessels containing more electrolyte solution. The electrodes are also dipped into these vessels as shown. The sample is spotted in the center of the sheet, and the oppositely charged ions then

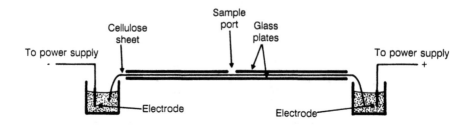

FIGURE 8.16   A paper electrophoresis apparatus.

have room to migrate in opposite directions on the sheet. Qualitative analysis is performed much as with paper chromatography by comparing the distances the individual components have migrated to that for standards spotted on the same sheet. (It may be necessary to render the spots visible prior to the analysis, as with paper chromatography.)

Problems associated with paper electrophoresis include the siphoning of electrolyte solution from one vessel through the paper to the other vessel when the levels of solution in the two vessels are different causing the spots to possibly migrate in the wrong direction. The solution to this problem is to ensure that the levels in the two vessels are the same. Another problem stems from the fact that oxidation/reduction processes are occurring at the surfaces of the electrodes. This may introduce undesirable contaminants to the electrolyte solution. These contaminants may in turn migrate onto the sheet. The solution to this problem is to isolate the electrodes while still allowing electrical contact, such as with the use of baffles, to keep the contaminants from diffusing from the vessels.

A typical gel electrophoresis apparatus is shown in Figure 8.17. The thin gel slab referred to above is contained between two glass plates. The slab is held in a vertical position and has notches at the top where the samples to be separated are spotted or "streaked." In the configuration shown in the figure, only downward movement takes place, and thus only one type of ion, cation or anion, can be separated, since there is only one direction to go from the notch.

A "tracking dye" can be added to the sample so that the analyst can know when the experiment is completed (the leading edge of the sample solvent is visible via the tracking dye). Also, the slab can be removed from the glass plates, and a staining dye can be applied which binds to the components, rendering them visible. The result is shown in Figure 8.18. Components with different mobilities through the gel show up as different

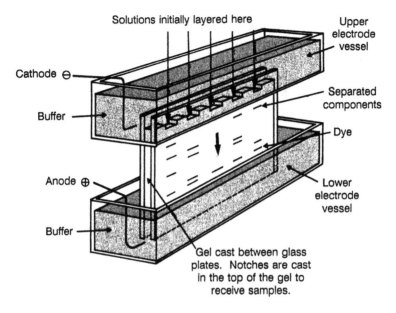

FIGURE 8.17   A gel electrophoresis apparatus (see text for description).

FIGURE 8.18   The appearance of the gel slab after the electrophoresis experiment and after the components are rendered visible via staining.

bands or streaks on the gel. Qualitative analysis is performed as with paper electrophoresis; standards are applied alongside the samples, and the components are identified by their positions relative to the standards.

Finally, isoelectric focusing has been a useful extension of basic gel electrophoresis in protein analysis. In this technique, a series of ampholytes is placed on the slab via electrophoresis. An ampholyte is a substance whose molecule contains both an acidic and basic functional group. Solutions of different ampholytes have different pH values. Different ampholyte molecules differ in size and therefore will have varying mobilities in the electrophoresis experiment. Thus, these molecules migrate into the slab and take up different positions along the height of the slab and create a pH gradient through the height of the slab. Amino acid molecules have different mobilities in different pH environments and also have their charges neutralized at particular pH values, rendering them immobile at some position in the gel. The pH at which the sample component is neutralized is called the "isoelectric point," and this technique is called "isoelectric focusing," since samples are separated according to their components' isoelectric points.

CHAPTER 9

GAS CHROMATOGRAPHY

9.1 INTRODUCTION

One instrumental chromatography configuration mentioned in the preceding chapter is one in which the mobile phase is gaseous and the stationary phase is either liquid or solid. This configuration is called gas chromatography and is abbreviated simply GC. It may also be abbreviated either GLC or GSC in order to stipulate whether the stationary phase is a liquid (GLC) or a solid (GSC). Most gas chromatography procedures utilize a liquid stationary phase (GLC), and thus the chromatography "type" (see Chapter 8 for the distinction between "type" and "configuration") is partition chromatography most of the time.

For the novice having just read Chapter 8, it may be difficult to visualize a chromatography procedure that utilizes a gas for a mobile phase. The gas, often called the "carrier gas," is typically purified helium or nitrogen. It flows from a compressed gas cylinder container via the regulated pressure of the cylinder and flow controller through a "column" containing the stationary phase where the separation takes place. There are two types of columns; two methods of holding the stationary phase in place. These will be discussed in Section 9.4.

The fact that the mobile phase is a gas creates additional unique features, one of which has to do with the mechanism of the separation.

When we discussed partition chromatography in Chapter 8, we were, for the most part, assuming a liquid mobile phase. The partitioning mechanism in that case involved only the relative solubilities of the mixture components in the two phases. With GLC, however, the mechanism, while still involving the solubilities of the mixture components in one phase, the liquid stationary phase, also involves the relative vapor pressures of the components, since we must have partitioning between one phase that is a liquid and another phase that is a gas. Further, since the mobile phase is a gas, the mixture components must also be gases, or at least liquids with relatively high vapor pressures (elevated temperatures are used), in order to be carried through the column as gases using a gaseous mobile phase.

Let us briefly review the concept of vapor pressure. Simply defined, vapor pressure is the pressure exerted by the vapors of a liquid above a liquid phase containing that liquid. It is a measure of the tendency of the molecules of a liquid substance to "escape" the liquid phase and become gaseous. If there is a strong such tendency, the vapor pressure is high (typical of nonpolar, low molecular weight liquids). If, however, there is a weak such tendency, then the vapor pressure is low (typical of highly polar and/or high molecular weight substances). Thus, example liquid substances with high vapor pressures are diethyl ether, acetone, and chloroform. Example liquid substances with low vapor pressures are water, high molecular weight alcohols, and aromatic halides. In GC, substances with high vapor pressures will be strongly influenced by the moving gaseous mobile phase and will emerge from the column quickly if their solubilities in the stationary phase are low. If their vapor pressures are high, but their solubilities in the stationary phase are also high, then they will emerge more slowly. If their vapor pressure is low, but they have a high solubility in the stationary phase, the time required for emergence from the column will be long. The time a given mixture component is retained by the stationary phase in the column from the time it is first introduced is called the "retention time." Figure 9.1 and Table 9.1 summarize the vapor pressure concepts discussed here.

9.2 INSTRUMENT DESIGN

Compared to the classical open-column configuration (Chapter 8), instrumental chromatography instruments are highly automated. For

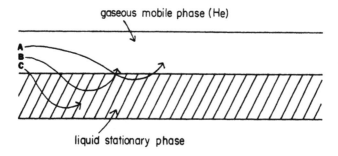

gaseous mobile phase (He)

liquid stationary phase

FIGURE 9.1   An illustration of the vapor pressure effects discussed in the text. Components A, B, and C have vapor pressures decreasing from A to B to C and have solubilities in the stationary phase increasing from A to B to C.

Table 9.1   A Summary of Retention Concepts for GC

Component's Vapor Pressure	Component's Solubility in Stationary Phase	Retention Time
High	Low	Short
High	High	Intermediate
Low	Low	Intermediate
Low	High	Long

example, the mobile phase is automatically and continuously fed into the column; the sample is introduced (injected) into the flowing mobile phase, which immediately carries it onto the column; and the components are automatically electronically detected as they elute from the column, greatly simplifying identification and quantitation. With GC, the instrument components include (1) the bottled compressed carrier gas, including the pressure regulator; (2) a flow controller at which the carrier gas flow rate is adjusted; (3) an injection port for introducing the sample; (4) the column with the stationary phase; (5) the detector; and (6) a recording system to accept the output of the detector. All of these components, except for the bottle of compressed carrier gas, are often found incorporated into a single unit. Perhaps most often, however, the recording system, which can be a simple strip-chart recorder (Chapter 5), is also a separate unit. A diagram of a gas chromatograph is shown in Figure 9.2.

    As mentioned earlier, the carrier gas is almost always helium or nitrogen. Helium is used most often. The regulated pressure from the bottled helium, typically 60–80 psi, is sufficient to sustain its flow through the system. Thus, a bottle of helium fitted with a regulator is almost

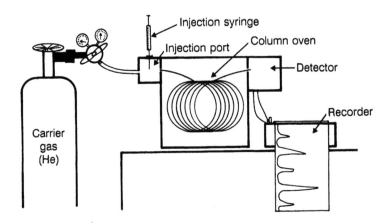

FIGURE 9.2–   A diagram of a gas chromatograph.

always seen standing next to the GC unit. The helium is often purified on-line with a molecular sieve column to remove traces of water prior to entering the instrument. It enters the instrument and flows through the flow controller, injection port, column, and detector before exiting into the laboratory air. Some laboratories in which GC is used extensively utilize small fume hoods to protect the quality of the laboratory air.

9.3 SAMPLE INJECTION

The injection port is designed to introduce samples quickly and efficiently. Most GC work involves the separation of volatile liquid mixtures. In this case, the injection port must be designed to flash vaporize small amounts of such samples so that the entire amount is immediately carried to the head of the column by the flowing helium. The most familiar design consists of a small glass-lined or metal chamber equipped with a rubber septum to accommodate injection with a syringe. This chamber is heated to a very high temperature (typically >200°C) and is connected directly to the column. As the helium "blows" through the chamber, a small volume of injected liquid (typically on the order of 0.1–3 μL) is thus flash vaporized and immediately carried onto the column. Small volume syringes with sharp, beveled tips for piercing the septum are needed. A variety of sizes and some additional features are available to make the injection easy and accurate. Syringes manufactured by the Hamilton Co., Reno, NV, are common. The mechanics for such an injection are shown in Figure 9.3. The rubber septum, after repeated sample introduction, can be replaced easily. Sample introduction systems for gases (gas sampling valves) and solids are also available.

A volume of liquid as small as 0.1–3 µL (appearing as an extremely small drop) may seem to be extraordinarily small. When vaporized, however, such a volume is much larger and will occupy an appropriate volume in the column. Also, the detection system, as we will see later, is very sensitive and will detect very small concentrations even in such a small volume. As a matter of fact, too large a volume is a concern to the operator, since columns, especially capillary columns, can become overloaded even with volumes that are very small. Overloading means that the entire vaporized sample will not fit onto the column all at once and will be introduced over a period of time. Obviously, a good separation would be in jeopardy in such an instance. To guard against overloading of capillary columns, split injectors have been developed. In these injectors, only a fraction of the liquid from the syringe actually is passed to the column. The remaining portion is split from the sample and vented to the air.

As implied above, the appropriate range of sample injection volume depends on column diameter. As we will see in the next section, column diameters vary from capillary size (0.2–0.3 mm) to $1/8$ and $1/4$ in. Table 9.2 gives the maximum injection volumes suggested for these column diameters. The capillary columns are those in which the overloading problem mentioned above is most relevant. Injectors preceding the $1/8$ in. or larger columns are not split.

The accuracy of the injection volume measurement can be very important for quantitation, since the amount of analyte measured by the detector depends on the concentration of the analyte in the sample as well as the amount injected. In Section 9.9, a technique known as the internal standard technique will be discussed. (It is also discussed in Chapter 5.) Use of this technique negates the need for superior accuracy with the injection volume, as we will see. However, the internal standard is not always used. Very careful measurement of the volume with the syringe in that case is paramount for accurate quantitation. An acceptable procedure for this is presented in Table 9.3. Of course, if a procedure calls only for identification (Section 9.8), then accuracy of injection volume is less important.

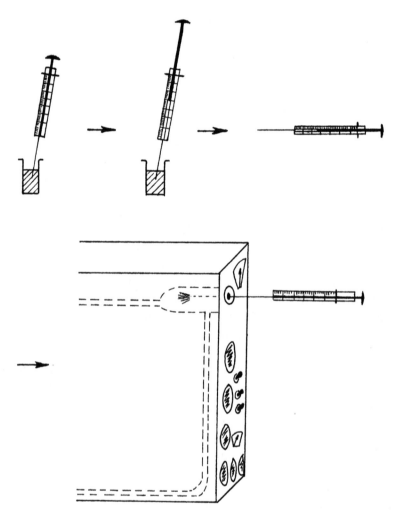

FIGURE 9.3   The mechanics of loading a GC syringe, showing the careful positioning of the plunger to the correct volume, and the introduction of the sample to the injection port.

9.4 COLUMNS

9.4.1 Instrument Logistics

GC columns, unlike any other type of chromatography column, are typically very long. Lengths varying from 2 ft up to 300 ft or more are

Table 9.2   Suggested Maximum Injection Volumes for Various Column Diameters

Column Diameters	Maximum Injection Volumes
¼ in. (packed column)	100 μL
⅛ in. (packed column)	20 μL
Capillary (open tubular)	0.1 μL

Table 9.3   A Syringe Loading and Injection Method When Accuracy of Injection Volume is Important

1. Flush the syringe thoroughly with clean sample solvent.
2. Expel the solvent from the syringe, then carefully retract the plunger (in air) to the 1.0 μL mark. A little less than 1 μL of solvent will be present (needle hold-up).
3. Transfer the syringe to the sample container and slowly draw several microliters of sample into the syringe barrel. Remove the syringe needle from the sample container and expel sample until the plunger is at the 2 μL mark (for a 1 μL injection).
4. Retract the syringe plunger, pulling the needle load entirely into the barrel. Two liquid plugs will be seen — sample and solvent without sample. Note the volume of the sample plug.
5. Insert the needle to its full length; inject the sample and quickly remove the syringe.

Source: Volume 2, Operators Manual for Varian Model 3300/3400 Gas Chromatograph. Used with permission of Varian Associates, Inc., Palo Alto, California.

possible. Additionally, it is important for the column to be kept at an elevated temperature during the run in order to prevent condensation of the sample components. Indeed, maintaining an elevated temperature is very important for other reasons, as we shall see in the next section. The obvious logistical problem is how to contain a column of such length and be able to simultaneously control its temperature.

Such a long column is wound into a coil and fits nicely into a small oven, perhaps 1–3 ft³ in size. This oven probably constitutes about half of the total size of the instrument (see Figure 9.4). Connections are made through the oven wall to the injection port and the detector. The temperatures of column ovens typically vary from 50 to 150°C, with higher temperatures possible in procedures that require them. A more thorough discussion of this subject is found in Section 9.5.

GC instruments are designed so that columns can be replaced easily by disconnecting a pair of brass fittings inside the oven. This not only facilitates changing to a different stationary phase altogether, but also allows the operator to replace a given column with a longer one contain-

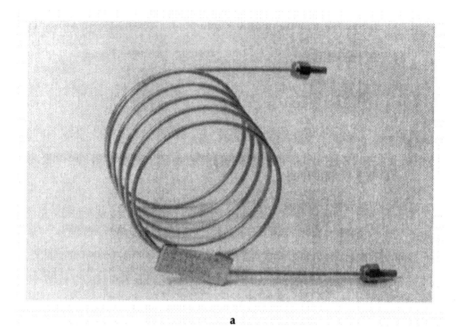

a

b

FIGURE 9.4   Examples of columns: (a) a 6-ft long, $1/8$ in. diameter packed column, and (b) a 100-ft long capillary column. Both are of such size as to fit into a small oven. (Courtesy of Varian Associates, Palo Alto, CA.)

ing the same stationary phase. The idea here is to allow more contact with the stationary phase, which in turn is bound to improve the separation. If a 6-ft column is useful for a partial separation, would not a 12-ft column be that much better?

9.4.2 Packed, Open Tubular, and Preparative Columns

It was indicated earlier that column lengths of up to 300 ft are not unusual. It should be mentioned here that the longer a column with stationary phase tightly packed (a so-called "packed" column), the greater the gas pressure required to sustain a flow. A 20-ft length is approximately the upper limit for the length of a "packed" column. A somewhat modern development in this area is the "open-tubular" capillary column. Instead of tightly packed solid substrate particles holding the liquid stationary phase inside the column (see Chapter 8, Section 8.6), the stationary phase is made to adsorb on the inside wall of a small diameter capillary tube so that the tube remains open to gas flow in the center. A design such as this offers very little resistance to gas flow and can be made hundreds of feet long without having to utilize a large pressure. It is no exaggeration to say that such columns are so popular today that the packed column is fast becoming obsolete (refer again to Figure 9.4 and to Figure 9.5).

In addition to the "analytical" columns (columns used mainly for analytical work), so-called "preparative" columns may also be encountered. Preparative columns are used when the purpose of the experiment is to prepare a pure sample of a particular substance (from a mixture containing the substance) by GC for use in other laboratory work. The procedure for this involves the individual condensation of the mixture components of interest in a cold trap as they pass from the detector and as their peak is being traced on the recorder. While analytical columns can be suitable for this, the amount of pure substance generated is typically very small, since what is being collected is only a fraction of the extremely small volume injected. Thus, columns manufactured with very large diameters (on the order of inches) and capable of very large injection volumes (on the order of milliliters) are manufactured for the preparative work. Also, the detector used must not destroy the sample, like the flame ionization detector (Section 9.7) does for example. Thus, the thermal conductivity detector (Section 9.7) is used most often with preparative GC.

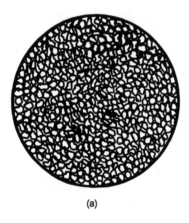 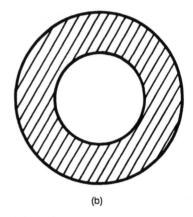

(a)   (b)

FIGURE 9.5   Illustrations of packed and capillary columns. (a) Cross-section of the packed column. (b) Cross-section of the open-tubular capillary column.

9.4.3 The Nature and Selection of the Stationary Phase

The liquid stationary phase in a GLC packed column is adsorbed on the surface of a solid substrate (also called the "support"). This material must be inert and finely divided (powdered). The typical diameter of a substrate particle is 125–250 μ, creating a 60- to 100-mesh material. These particles are of two general types: diatomaceous earth and Teflon. Diatomaceous earth, the decayed silica skeletons of algae, is most commonly referred to by the manufacturer's (Johns Manville's) trade name "Chromosorb." Various types of Chromosorb, which have had different pretreatment procedures applied, are available, such as Chromosorb P, Chromosorb W, and Chromosorb 101-104. The nature of the stationary phase, as well as the nature of the substrate material are usually specified in a chromatography literature procedure, and columns are tagged to indicate each of these as well.

Since the interaction of the mixture components with the liquid stationary phase plays the key role in the separation process, the nature of the stationary phase is obviously important. Several hundred different liquids useful as stationary phases are known. This means that the analyst has an awesome choice when it comes to selecting a stationary phase for a given separation. It is true, however, that relatively few such liquids

Table 9.4   Some Stationary Phases for GLC

Abbreviated or Nondescriptive Name	Structure, Descriptive Name, or Other Description	Useful for Mixtures of Compounds which are
FFAP	A Teflon-based material	Highly polar
Casterwax	$CH_3(CH_2)_5-CH-CH_2-CH=CH-(CH_2)_{17}-C$ with OH and an O (carbonyl) and OH group	Highly polar
Carbowax (variety of molecular weights)	$HO-[CH_2-CH_2-O]_n-H$	Polar
XE-60 (also XF1150, SF-1125)	$Si(CH_3)_3-O-[Si(CH_3)(CH_2-CH_2-CH_2-C\equiv N)-O]_n-Si(CH_3)_3$	Polar
OV-17	Methyl, phenyl, silicone (a silicone oil)	Somewhat polar
OV-101	Liquid methyl silicone	Nonpolar
OV-1	Methyl siloxane	Nonpolar
SE-30 (also DOW-200, DOW-11, SF-96)	$[-Si(CH_3)_2-O-Si(CH_3)_2-O-]_n$	Nonpolar
Apiezon (various types)	A grease	Nonpolar
Squalane	High molecular weight Hydrocarbon (C_{30})	Nonpolar

Source: "Basic Gas Chromatography" by McNair, H.M., and Bonelli, E.J., Varian Instruments, Walnut Creek, CA. 1969. Reprinted with permission, Varian Associates, Inc., Varian Analytical Instruments, Palo Alto, CA.

are in actual common use. Their composition is frequently not obvious to the analyst because a variety of common abbreviations have come to be popular for the names of some of them. Table 9.4 lists a number of common stationary phases, their abbreviated names, a description of their structures, and the classes of compounds (in terms of polarity) for which each is most useful.

The selection of a stationary phase depends largely on trial and error or experience, with consideration given to the polar nature of the mixture, as noted in Table 9.4 or a similar table. The usual procedure is to select a stationary phase, based on such literature information, and attempt the separation under the various conditions of column temperature, length, carrier gas flow rate, etc. to determine the optimum capability for separating the mixture in question. If this optimum resolution is not satisfactory (see Section 9.6), then an alternate selection is apparently required.

More experienced chromatographers may refer to the McReynolds Constants for a given stationary phase as a measure of its resolving power. A complete discussion of this subject, however, is beyond the scope of this text.

9.5 OTHER VARIABLE PARAMETERS

9.5.1 Column Temperature

Both the vapor pressure and the solubility of a substance in another substance change with temperature. Figure 9.6 shows, for example, how the vapor pressure of isobutyl alcohol and the solubility of acetanilide in ethanol change with temperature. It should not be surprising then that the precise control of the temperature of a GLC column is very important, since, as we have indicated, the separation depends on both vapor pressure and solubility. Both isothermal (constant) and programmed (continuously changing) temperature experiments are possible. For simple separations, the isothermal mode may well be sufficient; there may be sufficient differences in the mixture components' vapor pressures and solubilities to affect a good separation at the chosen temperature. However, for more complicated mixtures, a complete separation is less likely in the isothermal mode.

For example, consider gasoline which has a good number of highly volatile components, as well as a significant number of less volatile components. It is possible that at a temperature of, say, 100°C some of the less volatile components will be resolved, but the more volatile ones

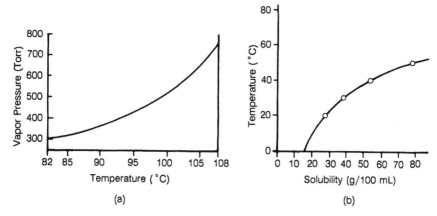

FIGURE 9.6   (a) A graph showing how the vapor pressure of isobutyl alcohol changes with temperature. (From Zubrick, J.W., *Organic Chemistry Survival Manual*, 2nd ed. Copyright © John Wiley & Sons Inc., New York, 1988. With permission.) (b) A graph showing how the solubility of acetanilide in ethanol changes with temperature. (From Moore, J., D. Dalrymple and O. Rodig, *Experimental Methods in Organic Chemistry*. Copyright © 1982 by Saunders College Publishing, reprinted by permission of the publisher.)

will pass through unresolved and have very short retention times. A lower temperature of, say, 40°C may cause complete resolution of these more volatile components, but would result in unwanted long retention times for the less volatile components and perhaps also result in poorly shaped peaks for these. If we could increase the temperature from 40 to 100°C or higher in the middle of the run, however, we could have the best of both worlds — complete resolution and reasonable retention times for all peaks. Thus, temperature programmable ovens have been developed and are now commonplace on virtually all modern GC units. Temperature programming can consist of simple programs, such as that suggested above — a single linear increase from a low temperature to a higher temperature — but it can also be more complex. For example, a chromatography researcher may find that several temperature increases, and perhaps even a decrease, must be used in some instances to affect an acceptable separation. Most modern GC units are capable of at least a slow temperature decrease in the middle of the run since they are equipped with venting fans that bring ambient air into the oven to cool it. Both a simple program and a more complex program are represented in Figure 9.7.

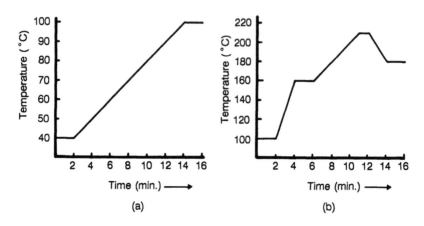

FIGURE 9.7   (a) A simple temperature program from 40 to 100°C at 5°C per minute. (b) A more complex temperature program.

9.5.2 Carrier Gas Flow Rate

The rate of flow of the carrier gas affects resolution. A simple analogy here will make the point. Wet laundry hung out on a clothesline to dry will dry faster if it is a windy day. The components of the mixture will "blow" through the column more quickly (regardless of the degree of interaction with the stationary phase) if the carrier gas flow rate is increased. Thus, a minimum flow rate is needed for maximum resolution. It is well known, however, that at extremely slow flow rates, resolution is dramatically reduced due to factors such as packing irregularities, particle size, column diameter, etc. The treatment of these factors and the quantitative determination of the optimum flow rate is beyond the scope of this text.

It is obvious that the flow rate must be precisely controlled. The pressure from the compressed gas cylinder of carrier gas, while sufficient to force the gas through a packed column, does not provide the needed flow control of itself. Thus a flow controller, or needle valve, must be part of the GC system and is often incorporated into the face of the instrument. In addition, the flow rate must be able to be carefully measured so that one can know what the optimum flow rate is and be able to match it in subsequent experiments. Various flow meters are commercially available for this and often the instrument manufacturer builds one into the instrument so that the flow rate is monitored continuously and is observable as one turns on the needle valve. In other cases, a simple soap bubble flow meter is often used and can be constructed easily from an old measuring pipet, a piece of glass tubing and a pipet bulb (see Figure 9.8). With this apparatus, a stopwatch is used to measure the time it takes a soap bubble squeezed from the bulb to move between to graduation lines, such as the 0 and 10 mL lines. The flow rate in milliliters per minute can thus be calculated.

9.6 THE CHROMATOGRAM

The chart recording giving the written record of the resolved substances, or peaks, is called the chromatogram. All qualitative and quantitative information obtained from a GC experiment is found in the chromatogram. One piece of such information is the "retention time," symbolized as t_R. From the time a substance is injected into the injection port until it emerges from the column and passes through the detector, it is being retained by the column. This is the span of time referred to as the retention time. Since the chart paper is passing through the recorder at a constant rate (for example, 1 in./min) the recorder itself becomes a device for measuring retention time. A certain number of inches or centimeters of chart paper corresponds to a certain number of minutes. Figure 9.9 shows how this measurement is made on the chromatogram. Typically, retention times vary from a small fraction of 1 min to about 20 min, although much longer retention times have been experienced.

Another parameter often measured is the adjusted retention time, t'_R. This is the difference between the retention time of a given component and the retention time, t_M, of an unretained substance, which is often air. You will recall the injection technique described in Table 9.3 involved the injection of some air with the sample. Air is usually completely unretained by a column, and thus the adjusted retention time becomes a measure of the exact time a mixture component spends in the stationary phase. Figure 9.10 shows how this measurement is made. The most important use of this retention time information is in peak identification or qualitative analysis. This subject will be discussed in more detail in Section 9.8.

Other parameters sometimes obtained from the chromatogram, which are mostly measures of column efficiency, are "resolution" (R), the number of "theoretical plates" (N), and the "height equivalent to a theoretical plate" (HETP or H). These require the measurement of the width of a

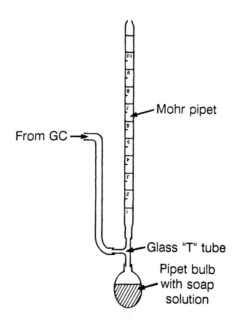

FIGURE 9.8   A homemade soap bubble flow meter constructed from an old Mohr pipet, a piece of glass tubing, and a pipet bulb.

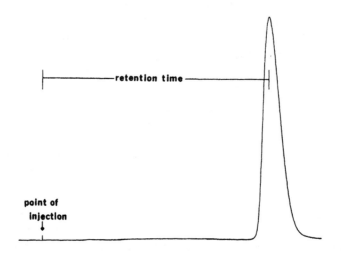

FIGURE 9.9   Retention time is the time corresponding to the length of chart paper measured from the point of injection to where the peak is at its apex.

peak at the peak base. This measurement is made by first drawing the tangents to the sides of the peaks and extending these to below the baseline, as shown for the two peaks in Figure 9.11. The width at peak

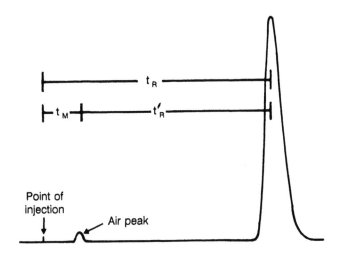

FIGURE 9.10   A chromatogram showing the definitions of t_R, t'_R, and t_M.

base, W_B, is then the distance between the intersections of the tangents with the baseline, as shown. Resolution is defined as the difference in the retention times of two closely spaced peaks divided by the average widths of these peaks, as shown mathematically in Figure 9.11. R values of 1.5 or more would indicate complete separation.

The number of theoretical plates, N, is also mathematically defined in Figure 9.11. The concept of theoretical plates was discussed briefly in Chapter 8, Section 8.3 for distillation. For distillation, one theoretical plate was defined as one evaporation/condensation step for the distilling liquid as it passes up a fractionating column. In chromatography, one theoretical plate is one "extraction" step along the path from injector to detector. You will recall in Chapter 8, Section 8.4 we spoke of chromatography as being analogous to a series of many extractions, but with one solvent (the mobile phase) constantly moving through the other solvent (the stationary phase), rather than being passed along through a series of separatory funnels. The equilibration that would occur in the fictional separatory funnel is one theoretical plate in chromatography.

The height equivalent to a theoretical plate, also mathematically defined in Figure 9.11, is that length of column that represents one theoretical plate or one equilibration step. Obviously, the smaller the value of this parameter, the more efficient the column. The more theoretical plates packed into a length of column the better the resolution. Factors that influence the number of theoretical plates and the resolution are column length, column temperature, carrier gas flow rate, and other

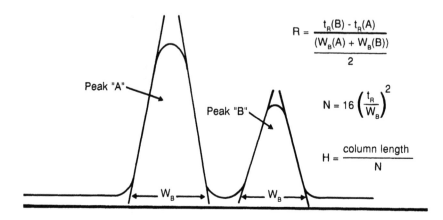

FIGURE 9.11   The measurement of the "width at base," which is needed for resolution and theoretical plate calculations.

factors we have already discussed. Other parameters calculated from the chromatogram, including capacity factor and selectivity, are defined in Chapter 10, Section 10.5.

9.7 DETECTORS

Detectors in GC are designed to generate an electronic signal when a gas other than the carrier gas elutes from the column. There have been a number of detectors invented to accomplish this. Not only do these detectors vary in design, but they also vary in sensitivity and selectivity. Sensitivity refers to the smallest quantity of mixture component for which it is able to generate an observable signal, and selectivity refers to the type of compound for which a signal can be generated. The flame ionization detector, for example, is a very sensitive detector, but does not detect everything, i.e., it is selective for only a certain class of compounds. The thermal conductivity detector, on the other hand, detects virtually everything, i.e., it is a "universal" detector, but is not very sensitive. What follows is a brief description of the designs of the detectors that are in common use, along with some indication of their sensitivity and selectivity.

9.7.1 Thermal Conductivity Detector (TCD)

The thermal conductivity detector (TCD) operates on the principle that gases eluting from the column have thermal conductivities different from that of the carrier gas, which is usually helium. Present in the flow channel at the end of the column is a hot filament, hot because it has an electrical current passing through it. This filament is cooled to an equilibrium temperature by the flowing helium, but is cooled differently by the mixture components as they elute, since their thermal conductivities are different from helium. This change in the cooling process causes the filament's electrical resistance to change and thus causes the current flowing through it and the voltage drop across it to change each time a mixture component elutes. The recorder, which is constantly monitoring this voltage drop, thus records a peak.

The actual design includes a second filament within the same detector block. This filament is present in a different flow channel, however, one through which only pure helium flows. Both filaments are part of a Wheatstone Bridge circuit as shown in Figure 9.12, which allows a "comparison" between the two resistances and a voltage output to the recorder as shown. Such a design is intended to minimize effects of flow rate, pressure, and line voltage variations.

Most recently, a flow modulated design has become popular. In this design, a single filament is used, and the column effluent is alternated with the pure helium through the flow channel where the filament is located. This eliminates the need to use two matched filaments.

The thermal conductivity detector is universal (detects everything), and it is nondestructive (can be used with preparative GC), but less sensitive than other detectors.

9.7.2 Flame Ionization Detector (FID)

Another very important GC detector design is the flame ionization detector (FID). In this detector, the column effluent is swept into a hydrogen flame where the flammable components are burned. In the burning process, a very small fraction of the molecules becomes fragmented, and

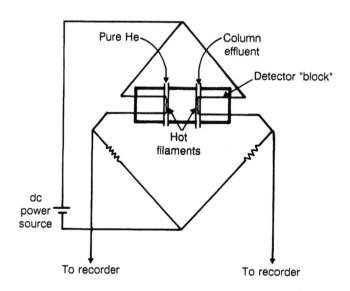

FIGURE 9.12   The thermal conductivity detector.

the resulting positively charged ions are drawn to a "collector" (negatively charged) electrode, a metal cylinder above and encircling the flame, while electrons flow to the positively charged burner head. The negatively charged collector and the positively charged burner head are part of an electrical circuit in which the current changes when this process occurs, and the change is amplified and seen as a peak on the recorder trace. Figure 9.13 shows a schematic diagram of this detector. The design includes the hydrogen flame burner nozzle, the collector electrode, an inlet for air to surround the flame, and an ignitor coil for igniting the hydrogen as it emerges from the nozzle.

Apparent on the exterior of the instrument and located near the bench on which the GC unit sits are pressure regulated compressed gas cylinders of hydrogen and air, as well as the helium. Metal tubing, typically $1/8$ in. diameter, connect the cylinders to the detector, with a needle valve for flow control in between. These valves are located in the instrument for easy access and control by the operator.

The FID is very sensitive, but is not universal, and also destroys (burns) the sample. It only detects organic substances that burn and fragment in a hydrogen flame. These facts preclude its use for preparative GC or for inorganic substances that do not burn, such as water, carbon dioxide, etc. Still, it is a very popular detector, given its sensitivity and given the fact that most analytical work involves flammable organic substances.

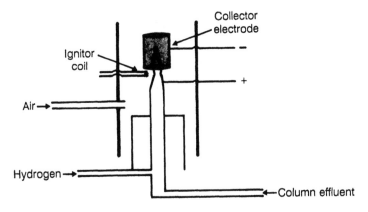

FIGURE 9.13   The flame ionization detector.

9.7.3 Electron Capture Detector (ECD)

A third type of detector, required for some environmental and bio-medical applications, is the electron capture detector (ECD). This detector is especially useful for large halogenated hydrocarbon molecules, since it is the only one which has an acceptable sensitivity for such molecules. Thus, it finds special utility in the analysis of halogenated pesticide residues found in environmental and biomedical samples.

The electron capture detector is another type of ionization detector. Specifically, it utilizes the beta emissions of a radioactive source, often nickel-63, to cause the ionization of the carrier gas molecules, thus generating electrons which constitute an electrical current. As an electrophilic component, such as a pesticide, from the separated mixture enters this detector, the electrons from the carrier gas ionization are "captured" creating an alteration in the current flow in an external circuit. This alteration is the source of the electrical signal which is amplified and sent on to the recorder. A schematic diagram of this detector is shown in Figure 9.14. The carrier gas for this detector is either pure nitrogen or a mixture of argon and methane.

An additional consideration regarding pesticides warrants mentioning here. Most of these compounds decompose on contact with hot metal surfaces. This problem has, however, been adequately solved for most pesticides by constructing the entire path of the sample out of glass or glass-lined materials. Thus, glass or glass-lined injection ports and all-glass columns are available.

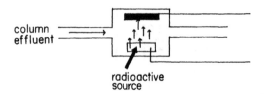

FIGURE 9.14   The electron capture detector.

In terms of advantages and disadvantages, the ECD is extremely sensitive, but only for a very select group of compounds — halogenated hydrocarbons. Other gases will not give a peak. It does not destroy the sample and thus may be used for preparative work.

9.7.4 Nitrogen/Phosphorus Detector (NPD)

While the ECD is useful for chlorinated hydrocarbon pesticides, the NPD, also known as the "thermionic" detector, is useful for the phosphorus and nitrogen-containing pesticides, the organophosphates and carbamates. The design, however, represents a slight alteration of the design of the FID. In the NPD, we basically have an FID with a bead of alkali metal salt positioned just above the flame. The hydrogen and air flow rates are lower than in the ordinary FID, and this minimizes the fragmentation of other organic compounds. These changes result in a somewhat mysterious increase in both the selectivity and sensitivity for the pesticides.

9.7.5 Flame Photometric Detector (FPD)

A detector that is specific for organic compounds containing sulfur or phosphorus is the flame photometric detector (FPD). A flame photometer (Chapter 7) is an instrument in which a sample solution is aspirated into a flame and the resulting emissions from the flame are measured with a phototube detector. The FPD is a flame photometer positioned to accept the effluent from the column in place of the aspirated sample. The flame in this case is a hydrogen flame as in the FID. The basic operating principle is that the sulfur or phosphorus compounds burn in the hydrogen flame and produce light emitting species. A monochromator, typically a glass filter, makes this detector specific for the compound of interest, and thus only one peak appears on the recorder trace. The signal for the recorder is the signal proportional to light intensity that is produced by the phototube.

The advantages are that it is a very selective detector and also very sensitive. Disadvantages include the problems associated with the need to carefully control the flame conditions so that the correct species are produced (S=S for the sulfur compounds and HPO for the phosphorus compounds). Such conditions include the gas flow rates and the flame temperature. It is a destructive detector.

9.7.6 Electrolytic Conductivity (Hall)

The Hall detector converts the eluting gaseous components into ions in liquid solution and then measures the electrolytic conductivity of the solution in a conductivity cell. The solvent is continuously flowing through the cell, and thus the conducting solution is in the cell for only a moment while the conductivity is measured and the peak recorded before it is swept away with fresh solvent. The conversion to ions is done by chemically oxidizing or reducing the components with a "reaction gas" in a small reaction chamber made of nickel positioned between the column and the cell. The nature of the reaction gas depends on what class of compounds is being determined. Organic halides, the most common application, use hydrogen gas at 850°C or higher as the reaction gas. The strong HX acids are produced, which give highly conductive liquid solutions.

The Hall detector has excellent sensitivity and selectivity, giving a peak for only those components which produce ions in the reaction chamber. It a destructive detector.

9.7.7 GC-MS and GC-IR

In Chapter 6, we discussed the fundamentals of mass spectrometry and IR spectrometry. The quadrupole mass spectrometer and the Fourier Transform IR spectrometer have been adapted to and used with GC equipment as detectors with great success in recent years. Gas chromatography-mass spectrometry (GC-MS) and gas chromatography-infrared spectrometry (GC-IR) are very powerful tools for qualitative analysis in GC because they not only give retention time information, but, due to their inherent speed, they are able to measure and record the mass or IR spectrum of the individual sample components as they elute from the GC column. It is like taking a photograph of each component as it elutes (see Figure 9.15). Coupled with the computer banks of mass and IR spectra, a component's identity is an easy chore for such a detector.

Recently fabricated GC-MS units have become very compact, in contrast to older units which take up large amounts of space. It seems the only disadvantage remaining is the expense, although that also seems to be improving. The only other slight disadvantage is the fact the large amounts of computer memory space are required to hold the amount of spectral information required for a good qualitative analysis.

Both the GC-MS instrument and the GC-IR instrument obviously require that the column effluent be fed into the detection path. For the IR instrument, this means that the IR cell, often referred to as a "light pipe," be situated just outside the interferometer (Chapter 6) in the path of the light, of course, but must also have a connection to the GC column and an exit tube where the sample may possibly be collected. The IR detector is nondestructive. With the mass spectrometer detector, we have the problem of the low pressure of the MS unit coupled to the ambient pressure of the GC column outlet. A special device is needed as a "go-between."

9.7.8 Photoionization Detector (PID)

The photoionization detector (PID), as the name implies, involves the ionization of eluting mixture components by light, specifically, UV light. The UV source emits a wavelength characteristic of the gas (either helium of argon) inside. This light passes into an "ionization chamber" through a metal fluoride window and into the path of the column effluent there. This is where the mixture components absorb the light and ionize. The resulting ions are detected through the use of a pair of electrodes in the ionization chamber, the current from which constitutes the signal to the recorder. The specific lamp and window are chosen according to the ionization energy needed for the compounds in the sample.

Since different lamps and windows are available, this detection method can often be selective for only some of the components present in the sample. Its sensitivity is especially good for aromatic hydrocarbons and inorganics. It is a very sensitive nondestructive detector.

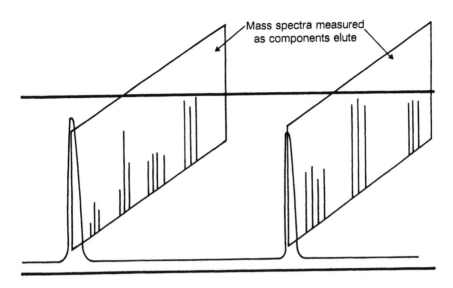

Mass spectra measured
as components elute

FIGURE 9.15   The "photographs" (the MS or IR spectra) of individual mixture components are obtained with GC-MS and GC-IR instruments.

9.8 QUALITATIVE ANALYSIS

As mentioned in Section 9.6, the parameters that are most important for a qualitative analysis using most GC detectors are the retention time, t_R, and the adjusted retention time, t'_R. Their definitions were graphically presented in Figure 9.10. Under a given set of conditions (the nature of the stationary phase, the column temperature, the carrier flow rate, the column length and diameter, and the instrument dead volume), the retention time is a particular value for each component. It changes only when one or more of the above parameters changes. Thus, repeated injections into a given system under a given set of conditions should always yield a particular retention time for a given compound, and qualitative analysis using this system only requires accurate measurement of t_R. When one of the parameters changes, such as when an analyst in another laboratory sets up with a different dead volume or perhaps a slightly different stationary phase composition, for example, then the retention time for that component will be slightly different. The adjusted retention time will correct for changes in the dead volume, but will not correct for any other change. A parameter defined as the "relative retention," however, will adjust for other changes. This parameter compares the retention of one component (1) with another (2) and is given the symbol alpha (α). It is defined as follows:

$$\alpha = \frac{t'_R(1)}{t'_R(2)} \qquad\qquad (9.1)$$

The relative retention is thus an important parameter for qualitative analysis if the work involves other setups with other instruments and columns which do not exactly match the original.

The usual qualitative analysis procedure, then, is to establish the conditions for the experiment, perhaps by trial and error in one's own laboratory or by matching conditions outlined in a given procedure, and then to match the retention time data, either ordinary retention time, t_R, or the relative retention, α, whichever is appropriate, for standards (pure samples) with that for the unknown. The analyst can then proceed to match the retention time data for the unknown to those of the pure samples to determine which substances are present.

One caution is that there may be more than one component with the same retention (no separation), and thus further experimentation may be required. For example, when working with a complex mixture whose components are perhaps not all known, it may be necessary to change the experimental conditions to determine whether a given peak is due to one component (known) or more (e.g., one known and one unknown). Changing the stationary phase may prove useful. Such a change would produce a chromatogram with completely different retention times and even possibly a different order of elution. Thus two components that were coeluted before may now be separated, evidence for which would be a different peak size for the known component.

9.9 QUANTITATIVE ANALYSIS

9.9.1 Peak Size Measurement

The physical size of the peak traced out by the strip chart recorder is directly proportional to the amount of that particular component passing through the detector. Thus, it is imperative that we have an accurate method for determining this peak size if it is the quantity of a component in the mixture that is sought. There are a variety of methods that have

height method, which simply measures the height from the baseline to the apex of the peak, the triangulation method, which measures the area of a triangle drawn to approximate the peak, and the half-width method, which measures the area of a rectangle drawn to approximate the peak. These three methods are illustrated in Figure 9.16.

While the first is a peak "height" method, the others are peak "area" methods. In the triangulation method, the height and base of the triangle is measured as shown and the area calculated (bh/2). In the half-width method, the height of the peak and the width of the peak at half-height are measured, and these represent the length and width of a rectangle, thus the area is again easily calculated (hw). None of these methods are terribly accurate, but they are fast and do not require expensive equipment. The peak height method is especially useful (fast) when only a rough indication of quantity is desired.

The most popular method of measuring peak size is by integration. Integration is an area measuring method in which a series of "heights" are measured from the moment the pen begins to deflect until the baseline is completely restored, as illustrated in Figure 9.17. This is conveniently done by computer, since a computer works with digital values derived from the analog data output by the instrument to the recorder (Chapter 5). The method is thus easy, fast, and accurate.

The computer hardware for integration can be one of a variety of designs, from a small unit designed only for measuring chromatography peaks (a "computing integrator") to a larger system, such as a microcomputer or other computer, programmed using independently prepared software. Figure 9.18 shows an example of a computing integrator and the printed output. Such a device often replaces the ordinary recorder, since it prints the peaks as a recorder would. It also records the retention time next to each peak as the peak is recorded. Note the area values in the sample printout (under the "area" heading). These values represent the sum of the series of digital values represented by the heights illustrated in Figure 9.17.

9.9.2 Quantitation Methods

Several different approaches exist as to what peaks are measured and how the mixture component of interest is actually quantitated. We now discuss two of the more popular methods (see also Chapter 5).

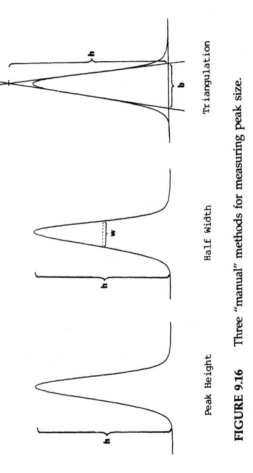

FIGURE 9.16   Three "manual" methods for measuring peak size.

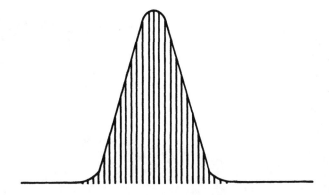

FIGURE 9.17   An illustration of the measurement of peak size by integration. The sum of a series of vertical "heights," such as illustrated, represents the area.

9.9.2a The Response Factor Method

Consider a four-component mixture to be analyzed by GC. The chromatogram may look something like that shown in Figure 9.19. One might think it logical that in order to quantitate the mixture for, say, component B, all one would need to do is to measure the sizes of all four peaks and divide the size of the peak representing B by the total of all four.

$$\%B = \frac{\text{Area}_B}{\text{Area}_A + \text{Area}_B + \text{Area}_C + \text{Area}_D} \tag{9.2}$$

The problems with this approach are (1) without comparing the peaks to a standard or a set of standards, it is not known whether the result is a weight percent, volume percent or mole percent. (2) The instrument detector does not respond to all components equally. For example, not all components will have the same thermal conductivity, and thus the thermal conductivity detector will not give equal sized peaks for equal concentrations of any two components. Thus, the sum of all four peaks would be a meaningless quantity, and the size of peak B by itself would not represent the correct fraction of the total.

It is possible, however, to measure a so-called response factor for the analyte, which is the area generated by a unit quantity injected, such as a microliter (μL) or microgram (μg). The procedure is to inject a known quantity of the analyte, measured by the position of the plunger in the

a

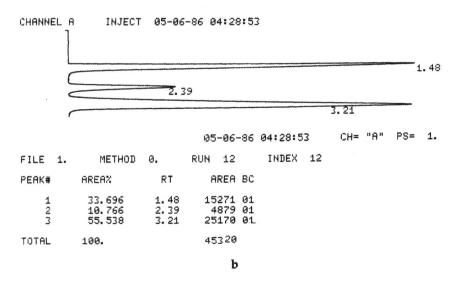

CHANNEL A   INJECT 05-06-86 04:28:53

1.48

2.39

3.21

05-06-86 04:28:53   CH= "A"   PS= 1.

FILE 1.   METHOD 0.   RUN 12   INDEX 12

PEAK#   AREA%   RT   AREA BC

1   33.696   1.48   15271 01
2   10.766   2.39   4879 01
3   55.538   3.21   25170 01L

TOTAL   100.   45320

b

FIGURE 9.18   (a) A computing integrator and (b) the printout from a computing integrator.

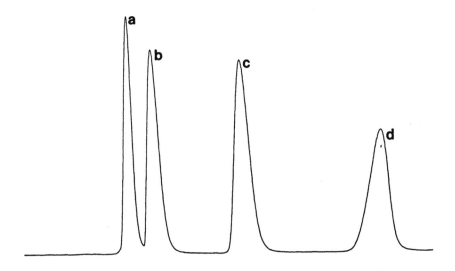

FIGURE 9.19   A chromatogram of a four-component mixture.

syringe (µL), or by weighing the syringe before and after filling. The peak size that results is measured and divided by this quantity:

$$\text{response factor} = \frac{\text{size of peak}}{\text{quantity of pure sample injected}} \qquad (9.3)$$

The quantity of analyte in an unknown sample is then determined by measuring the peak size of the analyte, resulting from an injection of a known quantity of unknown sample and dividing by the analyte's response factor:

$$\text{quantity of analyte} = \frac{\text{peak size}}{\text{response factor}} \qquad (9.4)$$

The percent of the analyte can then by calculated as follows:

$$\% \text{ of analyte} = \frac{\text{quantity of analyte} \left(\text{from Equation 9.4}\right)}{\text{total quantity injected}} \qquad (9.5)$$

In this method, only the peak of the analyte need be measured in the four-component mixture in order to quantitate this component.

9.9.2b Internal Standard Method

Since the peak size is directly proportional to concentration, one may think that one could prepare a series of standard solutions and obtain peak sizes for a plot of peak size vs concentration, a method similar to Beer's Law in spectrophotometry, for example. But since peak size also varies with amount injected, there can be considerable error due to the difficulty in injecting consistent volumes as discussed above and in Section 9.3. A method that does away with this problem is the internal standard method (see Chapter 5, Section 5.3). In this method, all standards and sample are spiked with a constant known amount of a substance to act as what is called an internal standard. The purpose of the internal standard is to serve as a reference point for the peak size measurements so that slight variations in injection technique and volume injected are compensated by the fact that the internal standard peak and the analyte peak are both affected by the slight variations, and thus the problem cancels out.

The procedure is to measure the peak sizes of both the internal standard peak and the analyte peak and then to divide the analyte peak size by the internal standard peak size. The "area ratio" thus determined is then plotted vs concentration of the analyte. The result is a method in which the volume injected is not as important and, in fact, can vary substantially from one injection to the next.

Can just any substance serve as an internal standard? There are certain characteristics that the internal standard should have, and these are listed below.

(1)   Its peak, like the analyte's, must be completely resolved from all other peaks.
(2)   Its retention time should be close to that of the analyte.
(3)   It should be structurally similar to the analyte.

9.10 TROUBLESHOOTING

Problems that arise during a GC experiment usually manifest themselves on the chromatogram. Examples of such manifestations are peak shapes being distorted, peak sizes diminishing for reasons other than quantity of analyte, the baseline drifting, the retention times changing for no apparent reason, etc. These kinds of problems can usually be traced to injection problems, problems with the column, or problems with the detector. There can, of course, be problems associated with the electronics of the instrument. However, we will not be concerned with those here because of the large number of different instrument designs that have been manufactured over the years. The operator can usually find assistance for these in a troubleshooting section of the manuals that accompany the instrument.*

In the following paragraphs, we will address some of the most common problems encountered, pinpoint possible causes, and suggest methods of solving the problems.

Diminished Peak Size — We could also refer to this as reduced sensitivity. The peaks are smaller than expected based on previous observations when equal or greater quantities of a particular sample were injected. Such an observation usually means a problem with injection (less injected than assumed) or a problem with the detector such that a smaller electronic signal is sent to the recorder. One should check for a leaky or plugged syringe, a worn septum, a leak in the pre- and post-column connections or a contaminated detector. Of course, detector attenuation, recorder sensitivity settings, electrical connections, and other associated hardware problems are potential causes.

Unsymmetrical Peak Shapes — Peak "fronting" or peak "tailing" (Figure 9.20) are typical examples of this problem. These could be indicators of poor injections, meaning too large an injection volume, too slow with the syringe manipulation during injection, or not fully penetrating the

---

*   See also the "GC Troubleshooting" column published regularly in the monthly journal
     LC•GC, The Magazine of Separation Science., Aster Publishing Corporation, Eugene, OR.

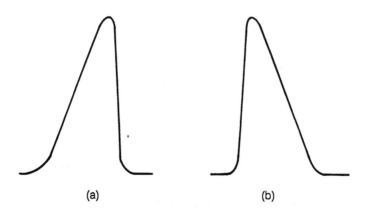

(a)   (b)

FIGURE 9.20   (a) A peak exhibiting fronting and (b) a peak exhibiting tailing.

septum. It may indicate a decomposition of thermal labile components in contact with the hot system components, such as the metal walls of the injection port and column. It may also mean contamination of the injection port and/or column.

Altered Retention Times — This is usually caused by changes in the carrier gas flow rate or column temperature. Flow rate changes can be caused by leaks in the system upstream from the column inlet, such as in the injection port (e.g., the septum); by low pressure in the system due to an empty or nearly empty carrier supply; or by faulty hardware, such as the flow control valve or pressure regulator. Temperature changes can be caused by a faulty temperature controller, an improperly set temperature program, too short a cool-down period prior to the next injection in a temperature programmed experiment, etc. This could also be caused by overloading the column or by diminished effectiveness (decomposition?) of the stationary phase.

Baseline Drift — This occurs when a new column has not been sufficiently conditioned, when the detector temperature has not reached its equilibrium value, or when the detector is contaminated or otherwise faulty. New columns need to be conditioned, usually with an overnight "bakeout" at the highest recommended temperature for that column. Detector signals may very well change when the detector temperature changes. One should be sure that sufficient time has been given for the detector temperature to level off. The nature of detector problems depends of course on the detector. TCD filaments may become oxidized due to an air leak, ionization detectors may be leaking, or there may be a crack in the FID burner nozzle, etc.

Baseline Perturbations — If the perturbations are in the form of spikes of an irregular nature, the problem is likely to be detector contamination. Such spikes are especially observed when dust particles have settled into the FID flame orifice. Of course, the problem may also be due to interference from electrical pulses from some other source nearby. Regular spikes can be due to condensation in the flow lines causing the carrier, or hydrogen (FID), to "pulse" or they can be due to a bubble flowmeter attached to the outlet of the TCD, as well as the electrical pulses referred to above. These can also be caused by pulses in the carrier flow due to a faulty flow valve or pressure regulator.

Appearance of Unexpected Peaks — Unexpected peaks can arise from components from a previous injection that moved slowly through the column; contamination from either the reagents used to prepare the sample or standards; or from a contaminated septum, carrier, or column. The solution to these problems include a rapid "bakeout" via temperature programming after the analyte peaks have eluted; using pure reagents; and replacing or cleaning septa, carrier, or column.

CHAPTER 10

HIGH PERFORMANCE LIQUID CHROMATOGRAPHY

10.1 INTRODUCTION

10.1.1 Basic Concepts

High performance liquid chromatography (HPLC) is an instrumental chromatography method in which the mobile phase is a liquid. The principles of liquid chromatography (LC) in general, including "types" or separation mechanisms, as well as a brief introduction to HPLC, are presented in Chapter 8. All types of liquid chromatography discussed in that chapter can be utilized in the HPLC configuration. Thus we have partition (LLC), adsorption (LSC), bonded-phase (BPC), ion-exchange (IEC and IC), and size-exclusion (SEC), including gel-permeation (GPC) and gel-filtration (GFC), all as commonly used types of HPLC. The instrumental design is based on concepts similar to gas chromatography (GC), especially in terms of the detection and recording schemes following the separation. You are referred to Chapter 9 for a discussion of GC principles.

Simply stated, HPLC involves the high pressure flow of a liquid mobile phase through a metal tube (column) containing the stationary phase, with electronic detection of mixture components occurring on the effluent end. The high pressure, often reaching 4000–6000 psi, is derived from a special pulsation-free pump, which will be described. The detection system can be any one of several designs, as with GC, and each of these will be discussed. In addition, special "solvent delivery" systems and injection systems are common and will also be described.

The rise in popularity of HPLC is due in large part to the advantages offered by this technique over the older, noninstrumental, "open column" method described in Chapter 8. The most obvious of these advantages is speed. Separation and quantitation procedures that require hours and sometimes days with the open column method can be completed in a matter of minutes, or even seconds, with HPLC. Modern column technology and gradient solvent elution systems, which will be described, have contributed significantly to this advantage in that extremely complex samples can be resolved with ease in a very short time.

The basic HPLC system, diagrammed in Figure 10.1, consists of a solvent (mobile phase) reservoir, pump, injection device, column, and detector. The pump draws the mobile phase from the reservoir and pumps it through the column as shown. At the head of the column is the injection device which introduces the sample to the system. On the effluent end, a detector, pictured in Figure 10.1 as a UV absorption detector, detects the sample components and the resulting signal is displayed as peaks on a strip-chart recorder. Besides these basic components, an HPLC unit may be equipped with a gradient programmer (Section 10.2), an auto sampler, a "guard column" and various in-line filters, and a computing integrator or other data handling system.

10.1.2 Comparisons with GC

There are many similarities between the HPLC configuration and the GC configuration. First, the stationary phase consists of small solid particles packed inside the column. Second, there is an injection device at the head of the column through which the mixture to be separated is introduced into the flowing mobile phase. Third, there is a detection system on the effluent side of the column which generates an electrical signal when something other than the mobile phase elutes. Fourth, the electronic

Concept of Liquid Chromatography

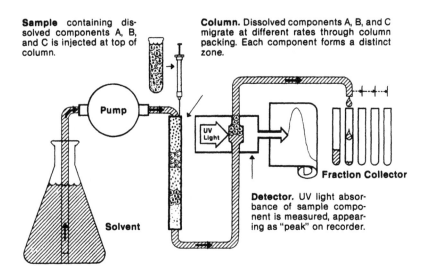

Sample containing dissolved components A, B, and C is injected at top of column.

Column. Dissolved components A, B, and C migrate at different rates through column packing. Each component forms a distinct zone.

Pump

UV Light

Fraction Collector

Detector. UV light absorbance of sample component is measured, appearing as "peak" on recorder.

Solvent

FIGURE 10.1   A diagram of an HPLC system in which mixture components A, B, and C are separated. (Courtesy of ISCO, Inc., Lincoln, NE.)

signal is fed into a strip-chart recorder where peaks are recorded — a system identical to GC.

Because the mobile phase is a liquid, however, there are also some very obvious differences in the two configurations. First, the mechanism of separation in HPLC involves the specific interaction of the mixture components with a specific mobile phase composition, while in GC, the vapor pressure of the components, and not their "interaction" with a specific carrier gas, is the most important consideration (see Chapter 9). Second, the force which sustains the flow of the mobile phase is that of a high pressure pump, rather than the regulated pressure from a compressed gas cylinder. Third, the injection device requires a totally different design due to the high pressure of the system and the possibility that a liquid mobile phase may chemically attack a rubber septum. Fourth, the detector requires a totally different design because the mobile phase is a liquid. Finally, the injector, column, and detector need not be heated as in GC, although the mode of separation occurring in the column can be affected by temperature changes, and thus sometimes elevated column temperatures are used.

10.1.3 Sample and Mobile Phase Pretreatment

The packed bed of finely divided stationary phase particles through which the mobile phase percolates is an excellent filter for the mobile phase and injected samples. Particles in the mobile phase as small as 5.0 $\times 10^{-6}$ cm in diameter can be filtered out by the stationary phase. The result of this is a decreased effectiveness of the column with time. In a reasonably short period of time, the particles filtered out on the column will (1) mask the stationary phase, preventing the mixture from interacting with it and thus causing poor resolution, and (2) make it necessary to use extremely high pressures in order to get the mobile phase through the column at the prescribed flow rate. The result is a dramatic decrease in the life of the column, an item of significant expense.

The solutions to this problem involve the prefiltering of all mobile phases and samples before beginning the experiment. For mobile phases and large sample volumes, this involves utilizing a vacuum apparatus, such as that pictured in Figure 10.2. There are many choices for filter materials. For nonaqueous solvents and their water solutions, paper is not a good choice due to the possibility of chemical attack which may cause contamination. For these, a Teflon-based, or other compatible material, should be used. A common Teflon-based designation is the PTFE designation. When these are used, if the mobile phase contains some water, the filter must be "wetted" first with some pure organic solvent in order to provide a reasonable filtration rate. Aqueous solutions are often impossible to filter unless the Teflon-based filter is first wetted in this manner. In cases in which only very small amounts of sample (or standard solutions) are available, a small syringe-type filtering unit is used. Here again, the filter must be wetted first if the filter is a Teflon-based material and the sample is partially or 100% aqueous.

In addition to these prefiltering steps, a series of in-line filters and a "guard column" are often used. The first such filter is at the very beginning of the mobile phase flow stream in the mobile phase reservoir, while others are at other strategic points, such as immediately following the injector. The guard column is usually placed just before the "analytical" column. Its function is often to remove not just particles, but other contaminating substances — substances that perhaps have long retention times on the analytical column and that eventually interfere with the detection in later experiments. Guard columns are inexpensive and disposable and are changed frequently.

FIGURE 10.2   A vacuum filtration apparatus for mobile phases and large volume samples. (Courtesy of Waters Division of Millipore, Inc., Milford, MA.)

Another problem is the appearance of air pockets in the HPLC system. If a sample or mobile phase has a significant amount of gas (air) dissolved in it, a pressure drop, which sometimes is experienced in an HPLC line, can cause these gases to withdraw from the solution. The air pockets that result are void spaces that can cause erratic readings from detectors, cause problems in pumps, and decrease the effectiveness of columns. The problem is alleviated by degassing the mobile phase and sample in advance of the their entering the system. Some instruments are equipped with degassing units between the mobile phase reservoirs and the pump. Usually, however, the analyst will degas the mobile phase and samples in a separate experiment in advance by creating a vacuum over the liquid with the use of a vacuum pump and/or agitating the liquid with use of an ultrasonic bath. A time-saving technique is to filter and degas in one step, since both procedures can involve the use of a vacuum. To do this, the vacuum (receiving) flask can be placed in the ultrasonic bath as the filtration proceeds.

10.2 SOLVENT DELIVERY

10.2.1 Pumps

The pump that is used in HPLC cannot be just any pump. It must a special pump that is capable of very high pressure in order to pump the mobile phase through the tightly packed stationary phase at a reasonable flow rate, usually between 0.5 and 4.0 mL/min. It also must be nearly free of pulsations so that the flow rate remains even and constant throughout. Only manufacturers of HPLC equipment manufacture such pumps.

There are several pump designs in common use. Probably the most common is the "reciprocating piston" pump shown in Figure 10.3. In this pump, a small piston is driven rapidly back and forth, drawing liquid in through the inlet check valve during its backward stroke and expelling the liquid through the outlet check valve during the forward stroke. Check valves allow liquid flow only in one direction. This design is often a "twin piston" design in which a second piston is 180° out of phase with the first. This means that when one piston is in its forward stroke, the other is in its backward stroke. The result is a flow that is free of pulsations. With the single piston design, a pulse damping device following the pump is desirable.

10.2.2 The Gradient Programmer

There are two mobile phase elution methods that are used to elute mixture components from the stationary phase. These are referred to as isocratic elution and gradient elution. Isocratic elution is a method in

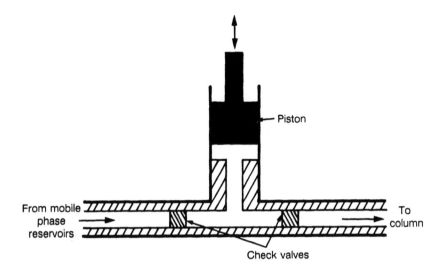

FIGURE 10.3   A diagram of a reciprocating piston HPLC pump.

which a single mobile phase composition is in use for the entire separation experiment. A different mobile phase composition can be used, but the transfer to a new composition can only be done by stopping the flow, changing the mobile phase reservoir, and restarting the flow. Gradient elution is a method in which the mobile phase composition is changed, often gradually, in the middle of the run, analogous to temperature programming in GC.

In any liquid chromatography experiment, the composition of the mobile phase is very important in the entire separation scheme. In Chapter 8, we discussed the role of a liquid mobile phase in terms of the solubility of the mixture components in both phases. Rapidly eluting components are highly soluble in the mobile phase and insoluble in the stationary phase. Slowly eluting components are less soluble in the mobile phase and more soluble in the stationary phase. Retention times, and therefore resolution, can be altered dramatically by a change in the mobile phase composition. The chromatographer can use this fact to his/her advantage by being able to change the mobile phase composition in the middle of the run. This is the basis for the gradient elution method.

The gradient programmer is a hardware module used to achieve this goal. The gradient programmer is capable of drawing from at least two mobile phase reservoirs at once and gradually, in a sequence programmed by the operator in advance, changing the composition of the mobile phase delivered to the HPLC pump. A schematic diagram of this system is shown in Figure 10.4a, and a sample "program" is shown in Figure 10.4b.

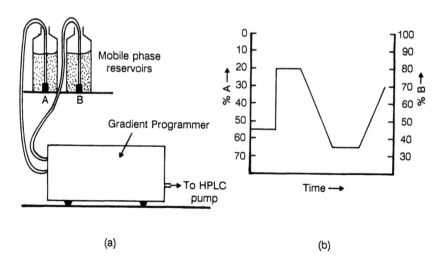

(a)   (b)

FIGURE 10.4   (a) A schematic diagram of a gradient programming system and (b) a sample program.

Solvent "strength" is a designation of the ability of a solvent (mobile phase) to elute mixture components. The greater the solvent strength, the shorter the retention times.

10.3 SAMPLE INJECTION

As mentioned previously, introducing the sample to the flowing mobile phase at the head of the column is a special problem in HPLC due to the high pressure of the system and the fact that the liquid mobile phase may chemically attack the rubber septum. For these reasons, the use of the so-called "loop injector" is the most common method for sample introduction.

The loop injector is a two position valve which directs the flow of the mobile phase along one of two different paths. One path is a sample loop, which when filled with the sample causes the sample to be swept into the column by the flowing mobile phase. The other path bypasses this loop while continuing on to the column, leaving the loop vented to the atmosphere and able to be loaded with the sample free of a pressure differential. Figure 10.5 is a diagram of this injector, showing both the "LOAD" position and the "INJECT" positions and the flow of the mobile phase in both positions. The sample loop has a particular volume which

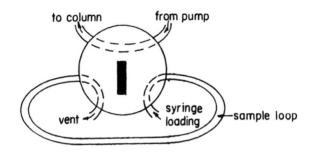

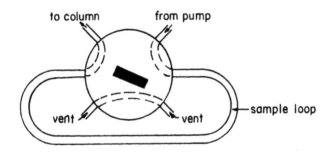

FIGURE 10.5   The loop injector for HPLC. (top) In the "LOAD" position, the sample is loaded into the loop via a syringe at atmospheric pressure. (bottom) In the "INJECT" position, the mobile phase sweeps the contents of the loop onto the column.

is of such accuracy that measuring the sample volume with the syringe loader is unnecessary, unless volumes smaller than this loop volume are required. This feature aids in the injection of an accurate, reproducible sample volume, which can increase the accuracy of a quantitation.

Automated injectors are often used when large numbers of samples are to be run. Most designs involve the use of the loop injector coupled to a robotic needle which draws the samples from vials arranged in a carousel-type auto sampler. Some designs even allow sample preparation schemes, such as extraction and derivatization, to occur prior to injection.

10.4 COLUMN SELECTION

The stationary phases available for HPLC are as numerous as those available for GC. As mentioned previously, however, adsorption, partition (including adsorbed and bonded-phase, BPC — see Chapter 8), ion-exchange, and size-exclusion are all LC methods. We can therefore at least classify the stationary phases according to which of these four types of chromatography they represent. Additionally, partition HPLC, which is the most common, is further classified as "normal-phase" HPLC or "reverse-phase" HPLC. Let us begin with these.

10.4.1 Normal-Phase Columns

Normal-phase HPLC consists of methods which utilize a nonpolar mobile phase in combination with a polar stationary phase. Adsorption HPLC actually fits this description, too, since the adsorbing solid stationary phase particles are very polar. (See discussion of adsorption columns in this section.) Normal-phase partition chromatography makes use of a polar liquid phase chemically bonded to these polar particles, which typically consists of silica, Si–O–, bonding sites. Sometimes called bonded-phase chromatography, this is also the method by which reverse-phase stationary phases and indeed some GC stationary phases are held in place. Typical examples of normal-phase bonded phases are those in which a cyano group (–CN), an amino group (–NH$_2$), or a diol group (–CHOH–CH$_2$OH) are part of the structure of the bonded-phase. Designation of such structural features are often given in the manufacturer's names. Some examples of these are Chromegabond DIOL, LiChrosorb DIOL, MicroPak-CN, uBondapak-CN, Nucleosil-NH$_2$, and Zorbax-NH$_2$. Typical mobile phases for normal-phase HPLC are hexane, cyclohexane, carbon tetrachloride, chloroform, benzene, and toluene.

10.4.2 Reverse-Phase Columns

Reverse-phase HPLC describes methods which utilize a polar mobile phase in combination with a nonpolar stationary phase. As stated earlier, the nonpolar stationary phase structure is a bonded-phase — a structure that is chemically bonded to the silica particles. Here, typical column names often have the carbon number designation indicating the length of a carbon chain to which the nonpolar nature is attributed. Typical designations are C$_8$ or C$_{18}$ (or ODS, meaning "octadecyl"), etc. Some of

Table 10.1   Some Adsorption HPLC Stationary Phases

Names	Types
Partisil	Silica, irregular
Hypersil	Silica, regular
Chromasep PAA	Alumina, irregular
Spherisorb A-Y	Alumina, regular
Corasil I and II	Silica, pellicular
Pellumina HS	Alumina, pellicular
Zipax	Silica, pellicular

these and other examples of reverse-phase stationary phases are Partisil ODS-2, uBondapak C_{18}, Spherisorb ODS, uBondapak Phenyl, Hyposil-SAS, and Nucleosil C-8. Common mobile phase liquids are water, methanol, acetonitrile (CH_3CN), and acetic acid buffered solutions.

10.4.3 Adsorption Columns

Adsorption HPLC is the classification in which the highly polar silica particles are exposed (no adsorbed or bonded liquid phase). Aluminum oxide particles fit this description too and are also readily available as the stationary phase. As mentioned earlier, this classification can also be thought of as normal-phase chromatography, but LSC rather than LLC. Typical normal-phase mobile phases (nonpolar) are used here. The stationary phase particles can be irregular, regular, or "pellicular" in which a solid core, such as a glass bead, is used to support a solid porous material. Examples of this classification are shown in Table 10.1.

10.4.4 Ion-Exchange and Size-Exclusion Columns

As discussed in Chapter 8, ion-exchange stationary phases consist of solid resin particles which have positive and/or negative ionic bonding sites on their surfaces at which ions are exchanged with the mobile phase (see Chapter 8, Figure 8.10). Cation exchange resins have negative sites so that cations are exchanged, while anion exchange resins have positive sites at which anions are exchanged. Typical examples are given in Table 10.2. A popular modern name for HPLC ion-exchange is simply "Ion

Table 10.2   Some Ion-Exchange Chromatography Stationary Phases

Names	Type
Ion-X-SC	Cation
Partisil 10 SCX	Cation
Amberlite IR-120	Cation
Dowex 50W	Cation
Ion-X-SA	Anion
Partisil 10 SAX	Anion
Amberlite IRA-400	Anion
Dowex 1	Anion

Chromatography." Detection of ions eluting from the HPLC column has posed special problems which are described in Section 10.6. The mobile phase for ion chromatography is always a pH-buffered water solution.

Size-exclusion columns, as discussed in Chapter 8, separate mixture components on the basis of size by the interaction of the molecules with various pore sizes on the surfaces of porous polymeric particles. Size-exclusion chromatography is subdivided into two classifications, gel-permeation chromatography (GPC) and gel-filtration chromatography (GFC). GPC utilizes nonpolar organic mobile phases, such as tetrahydrofuran (THF), trichlorobenzene, toluene, and chloroform, to analyze for organic polymers such as polystyrene. GFC utilizes mobile phases that are water-based solutions and is used to analyze for naturally occurring polymers, such as proteins and nucleic acids. GPC stationary phases are rigid gels, such as silica gel, whereas GFC stationary phases are soft gels, such as Sephadex. Neither technique utilizes gradient elution because the stationary phase pore sizes are sensitive to mobile phase changes.

10.4.5 Column Selection

Since each type of HPLC just discussed utilizes a different separation mechanism, the selection of a specific column packing (stationary phase) depends on whether or not the planned separation is possible or logical with a given mechanism. For example, if a given mixture consists of different molecules all of approximately the same size, then size-exclusion chromatography will not work. If a mixture consists only of ions, then ion chromatography is the logical choice. While the conclusions

Table 10.3   Summary of Applications of the Different Types of HPLC

Type	Useful for Components which
Normal and reverse-phase	Have a low formula weight (<2000)
	Are nonionic
	Are either polar or nonpolar
	Are water or organic soluble
Adsorption	Have a low formula weight (<2000)
	Are nonpolar
	Are organic soluble
Ion-exchange	Have a low formula weight (<2000)
	Are ionic
	Are water soluble
Size-exclusion	Have a high or low formula weight
	Are nonionic
	Are water or organic soluble

drawn from these examples are obvious, others are less obvious and require a study of the variables and the mechanisms in order to be able to logically choose a particular stationary phase.

Table 10.3 presents some guidelines about each choice which would be helpful in deciding which to use. While these guidelines may prove helpful as a starting point, additional facts about the planned separation need to be determined in order to select the most appropriate chromatographic system, including facts that can only be discovered through experimentation, or by searching the chemical literature. Several different mobile phase/stationary phase systems may work. Comparing reverse-phase with normal-phase, for example, one can see that there would only be a reversal in the order of elution. Polar components would elute first with reverse-phase, whereas nonpolar components would elute first with normal-phase. Experimenting with various mobile phase compositions, which may include a mixture of two or three solvents in various ratios, would be a logical starting point. Some considerations which would involve such experimentation are

1. The mixture components should have a relatively high affinity for the stationary phase compared to the mobile phase. This would mean longer retention times and thus probably better resolution.
2. The various separation parameters should be adjusted to provide optimum resolution. These include mobile phase flow rate, stationary phase particle size, gradient elution, and column temperature (using an optional column oven).

3. Use partition chromatography for highly polar mixtures and adsorption chromatography for very nonpolar mixtures.

10.5 THE CHROMATOGRAM

As with GC, the chart recording, which presents the written record of the separation, is called the chromatogram. Please refer to Chapter 9, Section 9.6 for a brief related discussion and for definitions of retention time (t_R), adjusted retention time (t'_R), resolution (R), the number of theoretical plates (N), and the height equivalent to a theoretical plate (H). In the HPLC definition of t'_R, the reference substance is not air, but the sample solvent, which usually gives a slight perturbation to the baseline at a very short retention time as it emerges from the column.

In addition to these parameters, liquid chromatographers are also concerned with the capacity factor, k', and selectivity, α. The capacity factor is the adjusted retention time divided by the retention time of the solvent, t_0.

$$k' = t'_R / t_0 \qquad (10.1)$$

The capacity factor is a measure of the retention of a component per column volume, since the retention time is referred to the time for the unretained solvent. The greater the capacity factor, the longer that component is retained and the better the chances for good resolution. This, however, must be weighed against the high speed advantage of HPLC. While a large capacity factor is desirable, the experiment should be completed within about 15 min. An optimum range for k' values is between 2 and 6.

Selectivity, α, is defined as the ratio of the adjusted retention time for one component to the adjusted retention time for another

$$\alpha = t'_R(A) / t'_R(B) \qquad (10.2)$$

and is a measure of the "quality" of a separation. Selectivity values greater than about 1.2 are considered good. A selectivity equal to 1 would mean that the two retention times are equal, which means no separation at all.

10.6 DETECTORS

The function of the HPLC detector is of course to examine the solution that elutes from the column and output an electronic signal proportional to the concentrations of individual components present there. In Chapter 9, we discussed a number of detector designs that serve this same purpose for GC. The design of the HPLC detectors, however, are more "conventional" in the sense that components present in a liquid solution can be determined with conventional instruments, such as a spectrophotometer. Thus, spectrophotometric and fluorometric detectors are common. Let us discuss some of the more popular HPLC detectors individually.

10.6.1 UV Absorption

The UV absorption HPLC detector is basically a UV spectrophotometer that is capable of measuring a flowing solution rather than a static solution. It has a light source, a wavelength selector, and a phototube as does an ordinary spectrophotometer. The sample compartment, however, is equipped with a "flow cell" through which the column effluent flows, and the absorbance is monitored continuously (see Figure 10.6). The output of the phototube is sent to the recorder or integrator where the absorbance is continuously displayed with time. Peaks are recorded as the UV absorbing components elute from the column.

The monochromator can be one of two different designs, either a simple light filter (a so-called fixed wavelength detector) or a full-blown slit/dispersing element/slit monochromator (a variable wavelength detector), which has a control on the face of the detector for dialing in the wavelength as in a standard spectrophotometer. The fixed wavelength detector can be made to be variable in the sense that the light filter can be changed, but one cannot tune to the wavelength of maximum absorbance, and thus some sensitivity can be lost. The fixed wavelength version, however, is less expensive and is fine for many applications for which UV absorbance detection is appropriate. Filters for 254 and 280 nm are common.

While such a detector is fairly sensitive, it is not universally applicable.

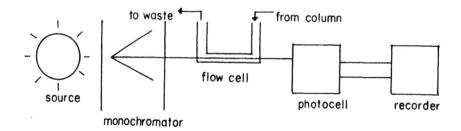

FIGURE 10.6   A schematic diagram of a UV absorbance detector. (From Kenkel, J., *Analytical Chemistry for Technicians*, Lewis Publishers, Inc., Chelsea, MI, 1988. With permission.)

The mixture components being measured must absorb light in the UV region and, at least in the fixed wavelength design, they must absorb at the wavelength used for a particular experiment in order for a peak to appear on the recorder. Also, the mobile phase must not absorb an appreciable amount at the selected wavelength.

10.6.2 Diode Array

A diode array UV detector is a "multichannel" detector in which the light beam from the UV source is not dispersed into its component wavelengths until after it has passed through the flow cell. The dispersed light then sprays across an array of photodiodes, each of which detects only a narrow wavelength band. With the help of a computer, the entire UV absorption spectrum can be immediately measured as each individual component elutes. With computer banks containing a library of UV absorption spectral information, a rapid, definitive qualitative analysis is possible in a manner similar to GCMS or GCIR (Chapter 9). In addition, the peak displayed on the recorder/integrator can be the result of a rapid changeover of the wavelength by the computer. Thus, the peaks displayed can represent the maximum possible sensitivity for each component. Finally, a diode array detector can be used to "clean up" a chromatogram so as to only display the peaks of interest. This is possible since we can rapidly change the wavelength giving rise to the peaks.

10.6.3 Fluorescence

The basic theory, principles, sensitivity, and application of fluorescence spectrometry (fluorometry) were discussed in Chapter 6. Like the UV absorption detector described above, the HPLC fluorescence detector is based on the design and application of its parent instrument, in this case the fluorometer, and you are referred to Chapter 6 for a review of the fundamentals of the fluorescence technique.

In summary, the basic fluorometer, and thus the basic fluorescence detector, consists of a light source and a monochromator (usually a filter) for creating and isolating a desired wavelength, a sample "compartment," and a second monochromator (another filter) with a phototube detector for isolating and measuring the fluorescence wavelength. The second monochromator and detector are lined up perpendicular to the light beam from the source (so-called "right angle" configuration).

As with the UV absorption detector, the sample compartment consists of a special cell for measuring a flowing, rather than static, solution. The fluorescence detector thus individually measures the fluorescence intensities of the mixture components as they elute from the column (see Figure 10.7). The electronic signal generated at the phototube is sent to the recorder or integrator where a peak is recorded each time a fluorescing species elutes.

The advantages and disadvantages of the fluorometry technique in general hold true here. The fluorescence detector is not universal (it will give a peak only for fluorescing species), but it is thus very selective (almost no possibility for interference) and very sensitive.

10.6.4 Refractive Index

The refractive index of a liquid or liquid solution is defined as the ratio of the speed of light in a vacuum to the speed of light in the liquid.

$$n = c_{vac}/c_{liq} \qquad (10.3)$$

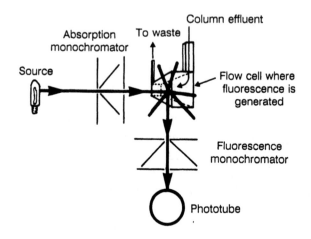

FIGURE 10.7   An illustration of a fluorescence detector.

Since the speed of light in any material medium is less than the speed of light in a vacuum, the numerical value of the refractive index for any liquid is greater than 1.

An instrument known as a refractometer has been invented and used for many years to measure the refractive index of liquids and liquid solutions for the purpose of both quantitative and qualitative analysis. A refractometer measures the degree of refraction (or "bending") of a light beam passing through a thin film of the liquid. This refraction occurs when the speed of light is different from a reference liquid or air. The refractometer measures the position of the light beam relative to the reference and is calibrated directly in refractive index values. It is rare for any two liquids to have the same refractive index, and thus this instrument has been used successfully for qualitative analyses.

The refractive index detector in HPLC is a modification of this basic instrument and actually can be purchased in two different designs, depending on the manufacturer. In probably the most popular design, both the column effluent and the pure mobile phase (acting as a reference) pass through adjacent flow cells in the detector. A light beam, passing through both cells, is focused onto a photosensitive surface, and the location of the beam when both cells contain pure mobile phase is taken as the reference point and the recorder pen is zeroed. When a mixture component elutes, the refractive index in one cell changes, and the light beam is "bent" and becomes focused onto a different point on the photosensitive surface, causing the recorder pen to deflect and trace a peak (see Figure 10.8).

The major advantage of this detector is that it is almost universal. All

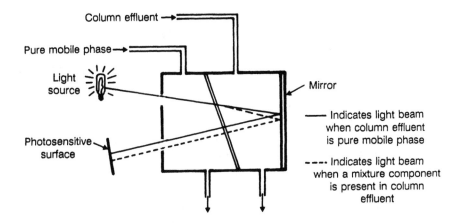

FIGURE 10.8   A representation of a retractive index detector (see text for description).

substances have their own characteristic refractive index (it is a physical property of the substance). Thus, the only time that a mixture component would not give a peak is when it has a refractive index equal to that of the mobile phase, a rare occurrence. The disadvantages are that it is not very sensitive and the output to the recorder is subject to temperature effects. Also, it is difficult to use this detector with the gradient elution method because it is sensitive to changes in the mobile phase composition.

10.6.5 Electrochemical

Various detectors which utilize electrical current or conductivity measurements for detecting eluting mixture components have been invented. These are called electrochemical detectors. Let us now examine some of the basic designs.

10.6.5a Conductivity

Perhaps the most important of all electrochemical detection schemes currently in use is the electrical conductivity detector. This detector is specifically useful for ion-exchange, or ion chromatography, in which the analyte is in ionic form. Such ions elute from the column and need to be detected as peaks on the recorder trace.

A well-known fact of fundamental solution science is that the presence of ions in any solution gives the solution a low electrical resistance and the ability to conduct an electrical current. The absence of ions means that the solution would not be conductive. Thus, solutions of ionic compounds and acids, especially strong acids, have a low electrical resistance and are conductive. This means that if a pair of conductive surfaces is immersed into the solution and connected to an electrical power source, such as a simple battery, a current can be detected flowing in the circuit. Alternatively, if the resistance of the solution between the electrodes is measured (with an ohmmeter), it would be low. Conductivity cells based on this simple design are in common use to determine the quality of deionized water, for example. Deionized water should have no ions dissolved in it and thus should have a very low conductivity. The conductivity detector is based on this simple apparatus.

For many years, the concept of the conductivity detector could not work, however. Ion chromatography experiments utilize solutions of high ion concentrations as the mobile phase. Thus, changes in conductivity due to eluting ions are not detectable above the already high conductivity of the mobile phase. This was true until the invention of so-called ion "suppressors." Today, conductivity detectors are used extensively in HPLC ion chromatography instruments which also include suppressors.

A suppressor is a short column (tube) that is inserted into the flow stream just after the analytical column. It is packed with an ion-exchange resin itself; a resin that removes mobile phase ions from the effluent, much like a deionizing cartridge removes the ions in laboratory tap water, and replaces them with molecular species. A popular "mixed bed" ion-exchange resin is used, for example, in deionizing cartridges, such that tap water ions are exchanged for H^+ and OH^- ions, which in turn react to form water. The resulting water is thus deionized. Of course, in the HPLC experiment, the analyte ions must *not* be removed in this process, and thus suppressors must be selective only for the mobile phase ions.

A typical design for a conductivity detector uses electrically isolated inlet and outlet tubes as the electrodes. This design is shown schematically in Figure 10.9.

10.6.5b Amperometric

A thorough discussion of electroanalytical techniques, including "polarography," "voltammetry," and "amperometry" is given in Chapter 11.

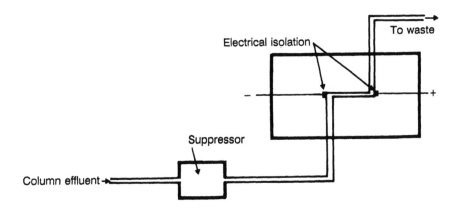

FIGURE 10.9   A drawing of a conductivity detector in which the inlet and outlet tubes are the electrodes.

An understanding of these would be useful for understanding the amperometric HPLC detector.

Electrochemical oxidation and/or reduction of eluting mixture components is the basis for amperometric electrochemical detectors. The three electrodes needed for the detection, the working ("indicator") electrode, reference electrode, and auxiliary electrode, are either inserted into the flow stream or imbedded in the wall of the flow stream (see Figure 10.10). The indicator electrode is typically glassy carbon, platinum or gold; the reference electrode is a silver/silver chloride electrode; and the auxiliary is a stainless steel electrode. Most often, the indicator electrode is polarized so as to cause oxidation of the mixture components as they elute. The oxidation current is then measured and constitutes the signal sent to the recorder/integrator.

Advantages of this detector include broad applicability to both ionic mixture components as well as molecular components, as long as they are able to be oxidized (or reduced) at fairly small voltage polarizations. Selectivity* can be improved by varying the potential. In addition, the sensitivity* experienced with this detector is quite good — generally better than the UV detector, but not as good as the fluorescence detector. A disadvantage is that the indicator electrode can become fouled due to products of the electrochemical reaction coating the electrode surface. Thus, this detector must be able to be disassembled and cleaned with relative ease, since this may need to be done frequently.

---

*   See Section 9.7 for definitions as related to chromatography detectors.

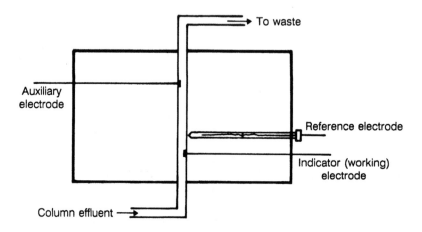

FIGURE 10.10 A drawing of an amperometric HPLC detector.

10.6.6 LC-MS and LC-IR

In Chapter 9, the use of mass spectrometry and FTIR for GC detection was discussed. Details of these techniques were individually given in Chapter 6. Much of the discussion presented in Chapter 9 is applicable here. Both mass spectrometry and infrared spectrometry have been adapted to HPLC detection in recent years.

FTIR is a "natural" for HPLC in that it is a technique that has been used mostly for liquids. The speed introduced by the Fourier Transform technique allows, as was mentioned for GC, the recording of the complete IR spectrum of mixture components as they elute, thus allowing the IR "photograph" to be taken and interpreted for qualitative analysis. Of course, the mobile phase, and its accompanying absorptions, is ever-present in such a technique and water must be absent if the NaCl windows are used, but IR holds great potential, at least for nonaqueous systems, as a detector for HPLC in the future.

The mass spectrometer is also incompatible with the HPLC system, but for a different reason. The ordinary mass spectrometer operates under very low pressure (a high vacuum — see Chapter 6), and thus the liquid detection path must rapidly convert from a very high pressure and large liquid volume to a very low pressure and a gaseous state. Several approaches to this problem have been used, but probably the most popular is the "thermospray" (TS) technique. In this technique, the column effluent is converted to a fine mist (spray) as it passes through a small-diameter heated nozzle. The analyte molecules, which must be thermally stable, are preionized with the presence of a dissolved salt. A portion of the spray is introduced into the mass spectrometer. The analyte and mobile phase must be polar if the TS technique is used because the mobile phase must dissolve the required salt and the components must interact with the analyte molecule.

10.7 QUALITATIVE AND QUANTITATIVE ANALYSIS

Qualitative and quantitative analysis with HPLC are very similar to that with GC (Chapter 9, Sections 9.8 and 9.9). In the absence of diode array, mass spectrometric, and FTIR detectors which give additional identification information, qualitative analysis depends solely on retention time data, t_R and t'_R. (Remember that t'_R here is the time from when the solvent front is evident to the peak). Under a given set of HPLC conditions — namely the mobile and stationary phase compositions, the mobile phase flow rate, the column length, temperature (when the optional column oven is used), and instrument dead volume — the retention time is a particular value for each component. It changes only when one of the above parameters changes. Refer to Section 9.8 for further discussion of qualitative analysis.

Peak size measurement and quantitation methods outlined for GC in Section 9.9 are also applicable here. The reproducibility of the amount injected is not nearly the problem with HPLC as it is with GC. Roughly ten times more sample is typically injected (5–20 μL), and there is no loss during the injection since the sample is not loaded into a higher pressure system through a septum. In addition, the sample loop is manufactured to have a particular volume and is often the means by which a consistent amount is injected, which means reproducibility is maximized through the consistent "overfilling" of the loop via the injection syringe. In this way, the loop is assured of being filled at each injection, and a reproducible volume is always introduced. Sometimes, however, the analyst chooses to inject varying volumes of a single standard to generate the standard curve (Chapter 5) rather than equal volumes of a series of standard solutions. In this case, the injection syringe is used to measure the volumes — a less accurate method, but better than an identical method with GC, since the sample volume is larger and there is less chance for sample loss.

With this type of quantitation, the standard curve is a plot of peak size vs amount injected, rather than concentration.

The most popular quantitation "method," then, is the series of standard solutions method with no internal standard (i.e., "serial dilution" — see Chapter 5) or the variable injection of a single standard solution as outlined above.

10.8 TROUBLESHOOTING

Problems that arise with HPLC experiments are usually associated with abnormally high or low pressures, system leaks, worn injector parts, air bubbles, and/or blocked in-line filters. Sometimes these manifest themselves on the chromatogram and sometimes they do not. In the following paragraphs, we will address some of the most common problems encountered, pinpoint possible causes, and suggest methods of solving the problems.*

Unusually High Pressure — A common cause of unusually high pressure is a plugged in-line filter. In-line filters are found at the very beginning of the flow line in the mobile phase reservoir, immediately before and/or after the injector and just ahead of the column. With time, they can become plugged due to particles that are filtered out (particles can appear in the mobile phase and sample even if they were filtered ahead of time), and thus the pressure required to sustain a given flow rate can become quite high. The solution to this problem is to backflush the filters with solvent and/or clean them with a nitric acid solution in an ultrasonic bath. Other causes of unusually high pressure are an injector blockage, mismatched mobile and stationary phases, and a flow rate that is simply too high. An injector that is left in a position between "LOAD" and "INJECT" can also cause a high pressure, since the pump is pumping, but their can be now flow.

Unusually Low Pressure — A sustained flow that is accompanied by low pressure may be indicative of a leak in the system. All joints should be checked for leaks (see next paragraph).

System Leaks — Leaks can occur within the pump, at the injector, at various fittings and joints, such as at the column, and in the detector.

---

*   See also the "LC Troubleshooting" column published regularly in the monthly journal *GC•LC, A Magazine of Separation Science.*, Aster Publishing Corporation, Eugene, OR.

Leaks within the pump can be due to failure of pump seals and diaphragms and loose fittings, such as at the check valves, etc. Leaky fittings should be checked for mismatched or stripped ferrules and threads, or perhaps they simply need tightening. Leaks in the injector can be due to a plugged internal line, or other system blockage, gasket failure, loose connections, or use of the wrong size syringe if the leak occurs as the sample is loaded. Detector leaks are most often due to a bad gasket seal or a broken flow cell. Of course, loose or damaged fittings and a blockage in the flow line beyond the detector are possible causes.

Air Bubbles — An air pocket in the pump can cause low or no pressure or flow, erratic pressure, and changes in retention time data. It may be necessary to bleed air from the pump or prime the pump according to system start-up procedures. Air pockets in the column will mean decreased contact with the stationary phase and thus shorter retention times and decreased resolution. Tailing and peak splitting on the chromatogram may also occur due to air in the column. Air bubbles in the detector flow cell are usually manifested on the chromatogram as small spikes due to the periodic interruption of the light beam (e.g., in a UV absorbance detector). Increasing the flow rate, or restricting and then releasing the post-detector flow, so as to increase the pressure, should cause such bubbles to be "blown" out.

Column "Channeling" — If the column packing becomes separated and a channel is formed in the stationary phase, the tailing and splitting of peaks will be observed on the chromatogram. In this case, the column needs to be replaced.

Recorder Zero vs Detector Zero — Both the detector and the recorder have "zeroing" capability. The recorder zero control is used when there is zero recorder input, such as when the input terminals are shorted. The detector zero control is used to zero the recorder pen when only the mobile phase is eluting. The two can be "mismatched." This problem is obvious on the chromatogram when changing the attenuation setting on the detector. At a less sensitive attenuation, the detector output may appear to be zero, but when the attenuation is changed to a more sensitive setting, there may be a sudden jump in the baseline, indicating that the detector output is actually not zero since the more sensitive setting is able to show a small offset from zero. The solution to this problem is to initially zero the detector at the more sensitive setting.

Decreased Retention Time — When retention times of mixture components decrease, there may be problems with either the mobile or stationary phases. It may be that the mobile phase composition was not restored after a gradient elution, or it may be that the stationary phase was altered due to irreversed adsorption of mixture components, or simply chemical decomposition. Use of guard columns (see previous discussion) may avoid stationary phase problems.

Baseline Drift — A common cause of baseline drift is a slow elution of adsorbed substances on the column. A column clean-up procedure may be in order, or it may need to be replaced. This problem may also be caused by temperature effects in the detector. Refractive index detectors are especially vulnerable to this. In addition, a contaminated detector can cause drift. The solution here may be to disassemble and clean the detector.

You are also referred to the troubleshooting guide in Chapter 9 (GC) for possible solutions to problems.

CHAPTER 11

ELECTROANALYTICAL METHODS

11.1 INTRODUCTION

The subject of electroanalytical chemistry encompasses all analytical techniques which are based on electrode potential and current measurements at the surfaces of electrodes immersed in the solution tested. Either an electrical current flowing between a pair of immersed electrodes or an electrical potential developed between a pair of immersed electrodes is measured and related to the concentration of some dissolved species.

Electroanalytical techniques are an extension of classical oxidation-reduction chemistry, and indeed oxidation and reduction processes occur at or within the two electrodes, oxidation at one and reduction at the other. Electrons are consumed by the reduction process at one electrode and generated by the oxidation process at the other. The complete system is often called a "cell," the individual electrodes "half-cells," and the individual oxidation and reduction reactions are the "half-reactions." Electrons flow on a conductor between the half-cells, and this flow constitutes the electrical current that is often measured. A "galvanic" cell is one in which this current flows spontaneously because of the strong tendency for the chemical species involved to give and take electrons. A

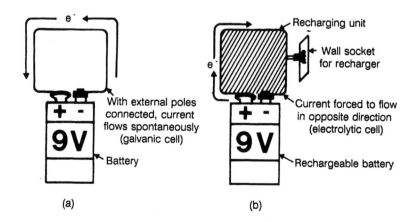

(a)   (b)

FIGURE 11.1   (a) A battery with its negative and positive poles connected is a galvanic cell. (b) A rechargeable battery, when positioned in its recharging unit, is an example of an electrolytic cell.

battery that has its positive and negative poles externally connected is an example of a such a cell. An "electrolytic" cell is one in which the current is not a spontaneous current, but rather is the result of connecting an external power source, such as a battery, to the system. A rechargeable battery, when it is positioned in the recharging unit, would be an example of such a cell (see Figure 11.1). Electroanalytical techniques utilize both general types of cells.

11.2 POTENTIOMETRY

Electroanalytical techniques which measure or monitor electrode potential utilize the galvanic cell concept. Such techniques fall under the general heading of "potentiometry." Examples include the pH measurement, ion-selective electrode measurement, and potentiometric titrations. In these techniques, a pair of electrodes is immersed, and the potential, or voltage, of one of the electrodes is measured, hence the name potentiometry. To understand how and why these techniques work, a fundamental knowledge of the Nernst Equation is needed.

11.2.1 The Nernst Equation

All oxidation and reduction half-reactions have a certain tendency to occur. Relative tendencies are often listed in text and reference books in the form of a table of "standard reduction potentials," symbolized E° and having the units of volts. If the tendency of a particular species to be reduced is high, then it will have a positive E°. For example, the reduction of fluorine, F_2 (to F^-), which is perhaps the species with the strongest of all tendencies to reduce, has an E° of $+2.87$ V. This represents what may be the upper limit to the E° values. If the tendency of a particular species to reduce is low, and in fact is more likely to be formed when another species is oxidized, then its E° value will be negative. Such is the case, for example, with all ions of alkali metals and alkaline earth metals. The E° value of lithium ion, Li^{+1} (being reduced to Li metal), is perhaps the most negative of all, -3.05. Thus the E° values for typical chemical species range from about -3 to about $+3$ V. These values are "relative" values, not absolute. This means that they are a measure of the relative tendency of each reaction to occur. They are based on the hydrogen ion reduction to hydrogen gas. This reaction is assigned an E° value of 0.00 V. Table 11.1 lists some half-reactions and accompanying E° values.

It is possible to determine the tendency, or E value, for an overall oxidation-reduction reaction. This is done by adding together the E° value for the reduction half-reaction and the $(-)E^\circ$ value for the oxidation half-reaction. The result is the "overall" oxidation-reduction reaction tendency, symbolized E°_{cell}. If the sign of this E° is positive, the reaction will proceed spontaneously to the right as written. If it is negative, it will proceed to the left.

Standard reduction and overall potentials are based on standard conditions of temperature (25°C), concentration of dissolved ionic species (1 M), and, when gases are involved, pressure (1 atm). If there is a deviation from these conditions, then the actual reduction and overall potential will be different from the standard. The relationship between the E value and these parameters is given by the Nernst Equation. For the general half-reaction

$$qQ^r + ne^- \rightarrow qQ^{r-n} \qquad (11.1)$$

Table 11.1 Standard Reduction Potentials for Selected Half-Reactions

Half-Reactions	E° (volts)
$F_2 + 2e^- \rightarrow 2F^-$	+2.87
$H_2O_2 + 2H^+ + 2e^- \rightarrow 2H_2O$	+2.07
$MnO_4^- + 8H^+ + 5e^- \rightarrow Mn^{+2} + 4H_2O$	+1.49
$Ce^{+4} + 1e^- \rightarrow Ce^{+3}$	+1.44
$Cl_2 + 2e^- \rightarrow 2Cl^-$	+1.36
$Cr_2O_7^{-2} + 14H^+ + 6e^- \rightarrow 2Cr^{+3} + 7H_2O$	+1.33
$O_2 + 4H^+ + 4e^- \rightarrow 2H_2O$	+1.23
$Hg^{+2} + 2e^- \rightarrow Hg$	+0.85
$Ag^+ + 1e^- \rightarrow Ag$	+0.80
$Fe^{+3} + 1e^- \rightarrow Fe^{+2}$	+0.77
$I_2 + 2e^- \rightarrow 2I^-$	+0.54
$Cu^{+2} + 2e^- \rightarrow Cu$	+0.34
$Hg_2Cl_2 + 2e^- \rightarrow 2Hg + 2Cl^-$ (SCE)	+0.24
$AgCl + 1e^- \rightarrow Ag + Cl^-$ (Ag/AgCl Ref.)	+0.22
$2H^+ + 2e^- \rightarrow H_2$	0.00
$Fe^{+3} + 3e^- \rightarrow Fe$	−0.04
$Fe^{+2} + 2e^- \rightarrow Fe$	−0.41
$Cr^{+3} + 3e^- \rightarrow Cr$	−0.74
$Zn^{+2} + 2e^- \rightarrow Zn$	−0.76
$Mg^{+2} + 2e^- \rightarrow Mg$	−2.38
$Na^+ + 1e^- \rightarrow Na$	−2.71
$K^+ + 1e^- \rightarrow K$	−2.92
$Li^+ + 1e^- \rightarrow Li$	−3.05

Reprinted with permission from the *CRC Handbook of Chemistry and Physics*, Weast, R., Ed., CRC Press, Inc., Boca Raton, FL.

the Nernst Equation* is

$$E = E^\circ - \frac{0.059}{n} \log \frac{\left[Q^{r-n}\right]^q}{\left[Q^r\right]^q} \qquad (11.2)$$

and for the general overall reaction

$$aA + bB \rightarrow cC + dD \qquad (11.3)$$

---

*   Strictly speaking, the Nernst Equation involves "activity" rather than concentration. Activity is directly proportional to concentration — the "activity coefficient" is the proportional constant. For most applications, the activity coefficient is equal to 1 and thus the activity equals the concentration. Further discussion of activity and activity coefficient is beyond the scope of this book.

the Nernst Equation is

$$E_{cell} = E^{\circ}{}_{cell} - \frac{0.059}{n} \log \frac{[C]^c [D]^d}{[A]^a [B]^b} \qquad (11.4)$$

If any species involved is a gas, the partial pressure of the gas is substituted for the concentration. If the temperature is different from 25°C, the "constant" 0.059 changes. The symbol "n" represents the number of electrons involved.

It is obvious from the foregoing discussion that the potential of electrodes and half-cells that one would actually measure is dependent on the concentration of the dissolved species involved. This is the basis for all quantitative potentiometry techniques and measurements to be discussed in this chapter.

11.2.2 Reference Electrodes

The measurement of any voltage is a relative measurement and requires an unchanging reference point. For voltage measurements in most electronic circuitry, this reference is usually "ground," which is often a wire that is connected to the frame of the electronic unit and the third prong in an electrical outlet, which in turn is connected to a rod that is pushed into the earth, hence the name "ground." Thus an electronics technician measures voltages relative to ground.

In electroanalytical chemistry, the unchanging reference is a portable half-cell that is designed to develop a potential that is constant. There is more than one design for this half-cell, and we now proceed to describe several that have become popular over the years.

11.2.2a The Saturated Calomel Reference Electrode (SCE)

The saturated calomel reference electrode is one such constant-potential electrode. A typical SCE available commercially is shown in Figure 11.2. It consists of two concentric glass tubes, which we will refer to here as an outer tube and an inner tube, each isolated from the other except for a small opening for electrical contact. The outer tube has a porous fiber

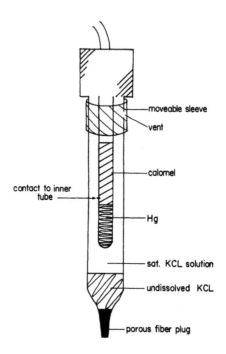

moveable sleeve

vent

calomel

contact to inner tube

Hg

sat. KCL solution

undissolved KCL

porous fiber plug

FIGURE 11.2   The saturated calomel reference electrode. (From Kenkel, J., *Analytical Chemistry for Technicians*, Lewis Publishers, Inc., Chelsea, MI, 1988. With permission.)

plug in the tip, which acts as the "salt bridge" to the analyte solution. A saturated solution of potassium chloride is in the outer tube. The saturation is evidenced by the fact that there is some undissolved KCl present. Within the inner tube is a paste-like material known as calomel. Calomel is made by thoroughly mixing mercury metal (Hg) with mercurous chloride (Hg_2Cl_2), a white solid. When in use, the following half-cell reaction occurs:

$$Hg_2Cl_2 + 2e^- \rightarrow 2Hg + 2Cl^-$$

The Nernst Equation for this reaction is:

$$E = E^\circ - \frac{0.059}{2} \log\left[Cl^-\right]^2$$

or

$$E = E^\circ - 0.059 \, \log\left[Cl^-\right]$$

Obviously the only parameter on which the potential depends is $[Cl^-]$. The saturated KCl present provides the $[Cl^-]$ for the reaction and, since it is a saturated solution, $[Cl^-]$ is a constant at a given temperature represented by the solubility of KCl at that temperature. If $[Cl^-]$ is constant, the potential of this half-cell, dependent only on the $[Cl^-]$, is also a constant. As long as the KCl is kept saturated and the temperature kept constant, the SCE is useful as a reference against which all other potential measurements can be made. Its standard reduction potential at 25°C (see Table 11.1) is +0.24 V. An advantage of this electrode is that, unlike its unsaturated counterparts, the $[Cl^-]$ does not change with evaporation of the water, since the solution remains saturated. Unsaturated calomel electrodes, however, are not affected by temperature changes like the SCE.

The SCE is usually used by dipping it into the analyte solution which is part of another half-cell, including an electrode typically referred to as an "indicator" electrode. A voltmeter is then externally connected across the leads to the two electrodes and the potential of the indicator electrode recorded vs the constant reference. While the SCE is dipped into the solution, there will be a slight leakage of the potassium and chloride ions through the porous tip. In order for the SCE to be used accurately, the experiment must not be adversely affected by the slight contamination from these ions. It is a good idea to slide the movable sleeve (Figure 11.2) downward so that the outer tube is vented while the electrode is in use so that the ions do indeed freely diffuse through the porous tip. Also, the electrode, under these circumstances, should not be immersed into the solution so deep that the level of solution in the external tube is lower than the level of the solution tested. This would cause the solution to diffuse into the SCE rather than the reverse, and thus would contaminate the KCl solution inside and possibly damage the SCE.

The vent hole may also be used prior to the experiment to refill the outer tube with more saturated KCl solution as this solution is lost with time. In addition, if the undissolved KCl disappears, more solid KCl can be added through the vent hole. The SCE should be stored in a beaker of distilled water (with the vent hole covered) for short-term storage and in a removable plastic cover usually provided with the electrode for long-term storage.

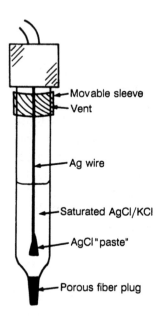

FIGURE 11.3 A typical silver-silver chloride reference electode.

11.2.2b The Silver-Silver Chloride Electrode

The commercial silver-silver chloride electrode is similar to the SCE in that it is enclosed in glass, has nearly the same size and shape, and has a porous fiber tip for contact with the external solution. Internally, however, it is different. There is only one glass tube and a solution saturated in silver chloride and potassium chloride is inside. A silver wire coated at the end with a silver chloride "paste" extends into this solution from the external lead (see Figure 11.3). The half-reaction which occurs is

$$AgCl(s) + 1e^- \rightarrow Ag(s) + Cl^-$$

and the Nernst Equation for this is

$$E = E^\circ - \frac{0.059}{1} \log[Cl^-]$$

The standard reduction potential for this half-reaction is +0.22 V. The

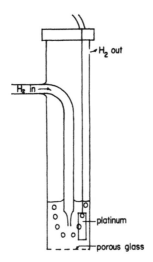

H₂ out

H₂ in →

platinum

porous glass

FIGURE 11.4   The standard hydrogen electrode. (From Kenkel, J., *Analytical Chemistry for Technicians*, Lewis Publishers, Inc., Chelsea, MI, 1988. With permission.)

potential is only dependent on the [Cl⁻], as was the SCE, and once again the [Cl⁻] is constant because the solution is saturated. Thus this electrode is also appropriate for use as a reference electrode. Refer to the segment above on the SCE for a discussion as to how it is physically handled while in use and in storage.

11.2.2c The Standard Hydrogen Electrode

The ultimate reference electrode utilizes the half-reaction on which Table 11.1 is based:

$$2H^+ + 2e^- \rightarrow H_2 \qquad\qquad E^\circ = 0.00 \text{ V}$$

This half-cell consists of a glass envelope housing a solution of constant pH in which a platinum wire or foil is immersed (see Figure 11.4). The platinum provides a surface for the exchange of electrons. Hydrogen gas bubbles through this solution at a constant pressure from some external source. Thus, since [H⁺] is a constant, and since the partial pressure of H_2 is a constant, the potential is a constant. The use of this cell as a reference electrode, however, is not very practical. It is cumbersome to use because it is not portable and requires a venting system for the hydrogen which otherwise is released into the laboratory.

11.2.3 Indicator Electrodes

As stated previously, the reference electrode represents half of the complete system for potentiometric measurements. The other half is the half at which the potential of analytical importance — the potential which is related to the concentration of the analyte — develops. There are a number of such "indicator" electrodes and analytical experiments which are of importance. Let us discuss these.

11.2.3a The pH Electrode

The measurement of pH is very important in many aspects of chemical analysis. Curiously, the measurement is based on the potential of a half-cell, the pH electrode.

The pH electrode, also called the glass electrode, consists of a closed-end glass tube that has a very thin fragile glass membrane at the tip. Inside the tube is a saturated solution of silver chloride that has a particular pH. It is typically a 1 M solution of HCl. A silver wire coated with silver chloride is dipped into this solution to just inside the thin membrane. While this is almost the same design as the silver-silver chloride reference electrode, the presence of the HCl and the fact that the tip is fragile glass and does not have a porous fiber plug points out the difference (see Figure 11.5).

The purpose of the silver-silver chloride combination is to prevent the potential that develops from changing due to possible changes in the interior of the electrode. The potential that develops is a "membrane" potential. Since the glass membrane at the tip is thin, a potential develops due to the fact that the chemical composition inside is different from the chemical composition outside. Specifically, it is the difference in the concentration of the hydrogen ions on opposite sides of the membrane that causes the potential — the membrane potential — to develop. There

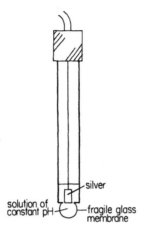

silver

solution of
constant pH

fragile glass
membrane

FIGURE 11.5   The pH or "glass" electrode. (From Kenkel, J., *Analytical Chemistry for Technicians*, Lewis Publishers, Inc., Chelsea, MI, 1988. With permission.)

is no half-cell reaction involved. The Nernst Equation is

$$E = E^{\circ} - 0.059 \log \frac{\left[H^{+}\right](\text{internal})}{\left[H^{+}\right](\text{external})}$$

or, since the internal $[H^{+}]$ is a constant, it can be lumped into E°, which is also a constant giving a modified E°, E^{*}, and eliminating $[H^{+}]$(internal):

$$E = E^{*} + 0.059 \log\left[H^{+}\right](\text{external})$$

In addition, we can recognize that $pH = -\log[H^{+}]$ and substitute this into the above equation:

$$E = E^{*} - 0.059 \, pH$$

The beauty of this electrode is that the measured potential (measured against a reference electrode) is thus directly proportional to the pH of the solution into which it is dipped. A specially designed voltmeter, called a pH meter, is used. In a pH meter, the meter readout is typically calibrated in both pH units and volts making the pH meter a rather versatile device.

The pH meter is standardized (calibrated) with the use of buffer so-lutions. Usually, two buffer solutions are used for maximum accuracy. The pH values for these solutions "bracket" the pH value expected for the sample. For example, if the pH of a sample to be measured is expected to be 9.0, buffers of pH = 7.0 and 10.0 are used. Buffers with pH values of 4.0, 7.0, and 10.0 are available commercially specifically for pH meter standardization. Alternatively, of course, homemade buffer solutions (see Chapter 3) may be used. In either case, the meter is adjusted, using the appropriate standardization method (refer to manufacturer's literature), to read the pH of the buffer into which the electrodes (the pH electrode and a reference electrode) are dipped. The solution of unknown pH is then determined. The pH electrode should be stored in a solution of distilled water or with a protective plastic sleeve over the tip.

11.2.3b The Combination pH Electrode

In order to use the pH electrode described above, two half-cells (some-times called "probes") are needed — the pH electrode itself and a ref-erence electrode, either the SCE or the silver-silver chloride electrode — and two connections are made to the pH meter. A modern development in this area is the invention of the "combination" pH electrode. This electrode incorporates both the reference probe and pH probe into a single probe. The reference portion is a silver-silver chloride reference. A drawing of the combination pH electrode is given in Figure 11.6.

The pH portion of this electrode is found in the center of the probe as shown. It is identical with the pH electrode described above — a silver wire coated with silver chloride immersed in a solution saturated with silver chloride and having a $[H^+]$ of $1.0\,M$. This solution is in contact with a thin glass membrane at the tip. The reference portion is in an outer tube concentric with the inner pH portion. It has a silver wire coated with silver chloride in contact with a solution saturated with silver chloride and potassium chloride. A porous fiber plug in the wall of the outer tube connects the outer tube with the solution tested as shown. This electrode either has two wires protruding from the lead (for connection to older style pH meters) or makes both connections using the "bnc" type or similar connector.

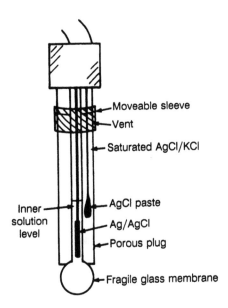

FIGURE 11.6   The combination pH electrode.

11.2.3c Ion-Selective Electrodes

The concept of the pH electrode has been extended to include other ions as well. Considerable research has gone into the development of these "ion-selective" electrodes over the years, especially in studying the composition of the membrane that separates the internal solution from the analyte solution. The internal solution must contain a constant concentration of the ion analyzed for, as with the pH electrode. Today we utilize electrodes with (1) glass membranes of varying compositions, (2) crystalline membranes, (3) liquid membranes, and (4) gas-permeable membranes. In each case, the interior of the electrode has a silver-silver chloride wire immersed in a solution of the analyte ion.

Examples of electrodes which utilize a glass membrane are those for lithium ions, sodium ions, potassium ions, and silver ions. Varying percentages of Al_2O_3, SiO_2, along with oxides of the metal analyte are often found in the membrane as well as other metal oxides. Selectivity and sensitivity of these electrodes vary.

With crystalline membranes, the membrane material is most often an

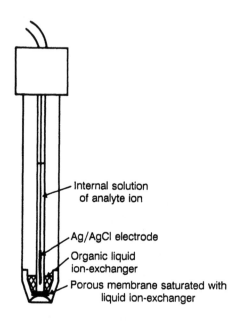

FIGURE 11.7   A liquid membrane electrode.

insoluble ionic crystal cut to a round flat shape and having a thickness of 1 or 2 mm and a diameter of about 10 mm. This flat "disc" is mounted into the end of a Teflon or PVC tube. The most important of the electrodes with crystalline membranes is the fluoride electrode. The membrane material for this electrode is lanthanum fluoride. The fluoride electrode is capable of accurately sensing fluoride ion concentrations over a broad range and to levels as low as 10^{-6} M. Other electrodes that utilize a crystalline membrane but with less impressive success records are chloride, bromide, iodide, cyanide, and sulfide electrodes. The main difficulty with these is problems with interferences.

Liquid membrane electrodes utilize porous polymer materials, such as PVC or other plastics. An organic liquid ion exchanger, immiscible with water, contacts and saturates the membrane from a reservoir around the outside of the tube containing the water solution of the analyte and the silver-silver chloride wire (see Figure 11.7). Important electrodes with this design are the calcium and nitrate ion-selective electrodes.

Finally, gas-permeable membranes are used in electrodes which are useful for dissolved gases, such as ammonia, carbon dioxide, and hydrogen cyanide. These membranes are permeated by the dissolved gases, but not by solvents or ionic solutes. Inside the electrode is a solution containing

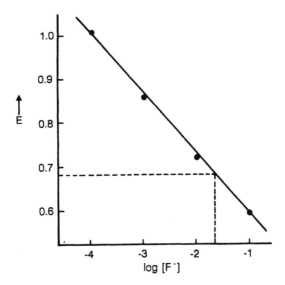

FIGURE 11.8   A calibration curve and unknown determination for an experiment using an ion-selective electrode.

the reference wire as well as a pH probe, the latter positioned so as to create a thin liquid film between the glass membrane of the pH probe and the gas-permeable membrane. As the gases diffuse in, the pH of the solution constituting the thin film changes, and thus the response of the pH electrode changes proportionally to the amount of gas diffusing in.

Calibration of ion-selective electrodes for use in quantitative analysis is usually done by preparing a series of standards as in most other instrumental analysis methods (see Chapter 5), since the measured potential is proportional to the logarithm of the concentration. The relationship is

$$E = E^* + \frac{0.059}{z} \log[\text{Ion}]$$

in which z is the signed charge on the ion. The analyst can measure the potential of the electrode immersed in each of the standards and the sample (vs the SCE or silver-silver chloride reference), plot E vs log[Ion], and find the unknown concentration from the graph. A typical plot and unknown determination is shown in Figure 11.8 for some real data using a fluoride ion-selective electrode. Alternatively, the standard additions technique (Chapter 5) may be used.

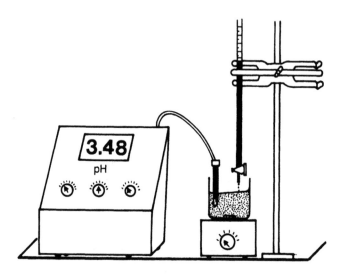

FIGURE 11.9   A potentiometric acid-base titration using a combination pH electrode.

11.2.4 Potentiometric Titrations

It is possible to monitor the course of a titration using potentiometric measurements. The pH electrode, for example, is appropriate for monitoring an acid-base titration and determining an end point in lieu of an indicator. The procedure has been called a "potentiometric titration," and the experimental setup is shown in Figure 11.9. The end point occurs when the measured pH undergoes a sharp change — when all the acid or base in the titration vessel is reacted. The same procedure can be used for any ion for which an ion-selective electrode has been fabricated and for which there exists an appropriate titrant.

In addition, potentiometric titration methods exist in which an electrode other than an ion-selective electrode is used. A simple platinum wire surface can be used as the indicator electrode when an oxidation-reduction reaction occurs in the titration vessel. An example is the reaction of Ce(IV) with Fe(II):

$$Ce^{+4} + Fe^{+2} \rightarrow Ce^{+3} + Fe^{+3}$$

If this reaction were to be set up as a titration, with Ce^{+4} as the titrant, the Fe^{+2} in the titration vessel, and the potential of a platinum electrode dipped into the solution monitored (vs a reference electrode) as the titrant is added, the potential would change with the volume of titrant added. This is because as the titrant is added, the measured E would change as the $[Fe^{+2}]$ is decreased, the $[Fe^{+3}]$ is increased, and the $[Ce^{+3}]$ is increased. At the end point and beyond, all the Fe^{+2} is consumed and the $[Fe^{+3}]$ and $[Ce^{+3}]$ change only by dilution, and thus the E is dependent mostly on the change in the $[Ce^{+4}]$. At the end point, there would be a sharp change in the measured E.

Automatic titrators have been invented which are based on these principles. A sharp change in a measured potential can be used as an electrical signal used to activate a solenoid and stop a titration (see also Chapter 12).

11.3 POLAROGRAPHY AND VOLTAMMETRY

11.3.1 Introduction

Techniques which utilize the electrolytic cell concept, rather than the galvanic cell concept as in potentiometry, are techniques which, as stated in Section 11.1, utilize an external power source to drive the cell reaction one way or the other. In these techniques, the current that results from this, and not the potential, is measured and related to the concentration of the analyte. Techniques in this category fall under the general heading of "amperometry" and include "polarography," which utilizes a special nonstationary mercury electrode to be described shortly, and "voltammetry," which utilizes only stationary electrodes.

Electrolytic cells can involve significantly larger currents than galvanic cells. The two-electrode system and the solution composition discussed for potentiometry must usually be modified in order to obtain desirable results. A reference electrode, such as the SCE or the silver-silver chloride electrode, cannot handle larger currents. In addition, the solution into which the electrodes are dipped must be able to sustain such a current. This was not true with potentiometric methods. Thus, for amperometric methods, a three-electrode system is usually used, and a "supporting electrolyte" must be added to the solution.

11.3.1a The Three Electrode System

The current measured is usually the current due to a specific oxidation or reduction process involving the analyte species at the surface of one of the electrodes. A three-electrode system includes a "working" electrode, at which the oxidation or reduction process of interest occurs, a reference electrode, such as the SCE or the silver-silver chloride electrode, and an "auxiliary" or "counter" electrode, which carries the bulk of the current (instead of the reference electrode) and "counters" the process that occurs at the working electrode. The three electrodes are connected to the power source which is a specially designed circuit for precise control of the potential applied to the working electrode and is often called a "potentiostat" or "polarograph." The working electrode can be made positive or negative with the flip of a polarity switch on the instrument. The use of the reference electrode is crucial for the precise control of the potential of the working electrode. The process occurring at the auxiliary electrode is unimportant analytically. However, it is important not to allow the products of the reaction occurring there to interfere with the process occurring at the working electrode. For this reason, the auxiliary electrode is often placed in a separate chamber with a fritted glass disc allowing electrical contact with the rest of the cell, but not allowing diffusion of undesirable chemical species to the working electrode (see Figure 11.10).

The typical experiment consists of scanning a potential range within which the process of interest is forced to occur at the surface of the working electrode. The current that flows as a result is measured by the instrument and displayed, usually on a recorder. This visual display has characteristics which depend on the particular potential application used, as we will see, and also on whether the electrode is stationary or nonstationary. The current due to the process of interest is determined from this display and related to the concentration of the analyte species.

11.3.1b The Supporting Electrolyte

The need for the supporting electrolyte mentioned above is obvious from the standpoint that the solution tested must be an electrolyte solution if it is to sustain a measurable current. Real-world sample solutions and standards which have very low analyte concentrations often do not have

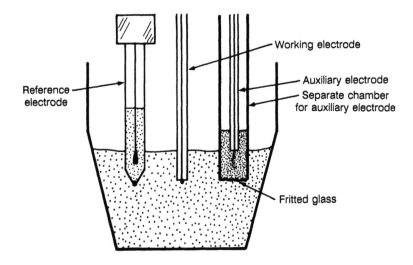

FIGURE 11.10   A drawing of a three-electrode system typical of most amperometric experiments.

a sufficient electrolyte content. In addition, the presence of a supporting electrolyte can help bring the potential of the electrode process into the desired range and free from interferences because of the possibility of complexation and other reactions between the analyte species and the supporting electrolyte species. Such a reaction can create a complex ion, for example, which could shift the potential required to an interference-free range. The possibility of the supporting electrolyte species itself interfering with the oxidation or reduction of the analyte is eliminated by choosing an electrolyte species that is difficult to oxidize or reduce compared to the analyte species or the complex ion. The instrument is capable of precise control of the potential, and thus capable of differentiating between the different species present in the solution. Thus, the analyte species can be "electroactive" (oxidized or reduced at the electrode surface) when the supporting electrolyte is not.

11.3.2 Polarography

11.3.2a Introduction

We have previously defined polarography as an amperometric technique which utilizes a special nonstationary electrode for the working electrode. We now proceed to describe this electrode in detail and continue with a discussion of the modern techniques associated with it.

The electrode is a "dropping mercury electrode" or DME. It consists of liquid mercury flowing through a very narrow-bore capillary tube, as shown in Figure 11.11. The mercury flows through the capillary by gravity from a reservoir of mercury. Drops of mercury form at the tip of the capillary, grow, fall, and reform in fairly rapid intervals (0.5–10 sec). This sequence of events is depicted in Figure 11.11. Figure 11.12 shows the complete apparatus with connections made from the electrodes to the instrument.

The DME was invented in the early 1900s by a Nobel Prize-winning chemist by the name of Heyerovsky. Its fluid nature eliminates two problems normally associated with amperometric measurements: (1) contamination of the electrode surface resulting from electrode reactions (the mercury surface is continuously renewed in the case of the DME), and (2) the decay of the current to small values with time due to the diffusion limited transportation of analyte to the electrode surface when the solution is not stirred (the solution *is* stirred in the case of the DME each time the drop falls).

As we will soon see in the sections to come, it is desirable for the instrument to know the precise moment at which the drop will fall. Mechanical drop timers have been invented which "knock" the capillary at regular intervals such that the drop falls when required. Modern drop timers utilize a larger bore capillary and a spring-loaded polyurethane "plug" at the top of the capillary for controlling the drop. Figure 11.13 shows this more recently developed unit. In either case, the precise time the drop is to fall is controlled electrically such that the instrument makes its measurement at that exact moment, usually just before the drop falls.

11.3.2b Classical Polarography

The "classical" or "normal" polarography experiment consists of scanning the potential range in which the analyte becomes electroactive and monitoring the current continuously at the same time. Since the current that will flow at any electrode surface is proportional to the area of the electrode surface exposed to the solution (in addition to the analyte concentration), the current vs potential display on the recorder shows a considerably unstable (but uniform) current. As the drop grows, the

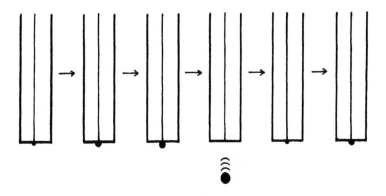

FIGURE 11.11   The dropping mercury electrode (DME) showing the growth and fall, etc. of the mercury drop. (From Kenkel, J., *Analytical Chemistry for Technicians*, Lewis Publishers, Inc., Chelsea, MI, 1988. With permission.)

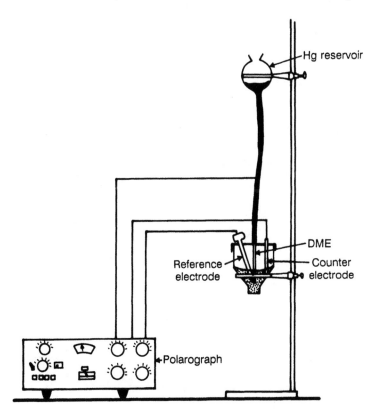

FIGURE 11.12   The "complete" polarography apparatus.

FIGURE 11.13   A modern apparatus for the mechanical control of the drop time and size. (Courtesy EG&G Princeton Applied Research, Princeton, NJ.)

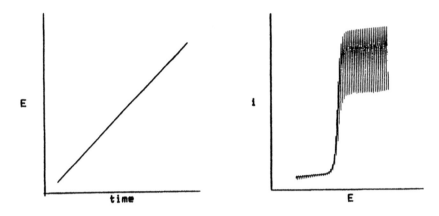

FIGURE 11.14   The E/time characteristic (left) and the recorder trace (right) for the classical polarography experiment described in the text. The "up and down" pattern on the right is due to the continuous growth and fall of the mecury drop.

current increases, but when the drop falls, the current drops. The continuous growing and falling of the drop is thus accompanied by a continuous up and down motion of the recorder pen. The potential-time characteristic and the resulting current display (polarogram) for the reduction of cadmium ion in 0.1 M NaNO$_3$ is shown in Figure 11.14. The potential-time characteristic is often called a "ramp" because of the linear increase apparent in the figure.

Parameters associated with the polarogram are the "diffusion current," i_d, the "half-wave potential," $E_{1/2}$, and the "background current." The diffusion current is the current which is due directly to the presence of the analyte species in the solution. It is the increase in the measured current that occurs when the potential for the reduction or oxidation of the analyte species is reached as electrode potential is linearly increased. The half-wave potential is the potential at which this increase in the current is half of its full value. The background current is the current that would be observed if the analyte species were not present in the solution. Sometimes this background current includes the current due to other oxidation or reduction processes occurring in the event there are other species in solution that can be reduced or oxidized at lower potentials. In the absence of these other species, this current would be very small and is called the "residual" or "charging" current. The diffusion current, half-wave potential, and residual current are all graphically defined in Figure 11.15.

The half-wave potential is often taken to be the potential at which

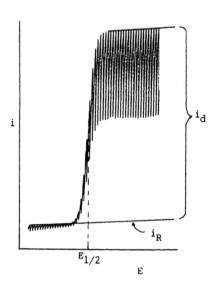

FIGURE 11.15   A classical polarogram and the definitions of diffusion current, i_d, half-wave potential, $E_{1/2}$, and residual current, i_R.

reduction or oxidation of a given species occurs. It can be thought of as the "barrier" to be crossed in order for oxidation or reduction to occur. This barrier can be crossed from either direction, but must be crossed in order to observe the process desired. Usually we cross it from one direction only and observe the magnitude of the current that results. In a few techniques, however, it is desirable to cross it while proceeding in one direction, then reverse the scan and cross it from the other direction too. We will describe these techniques in the next several subsections. Probably the most notable fact about the half-wave potential is that it is the one parameter which identifies a current increase to be due to a particular dissolved chemical species. One might think of it then as a parameter for qualitative analysis. However, the greatest usefulness of this lies in the fact that one can use it to identify the precise potential at which the current due to the analyte is expected and found and thus allow the correct diffusion current to be measured. This is important because it is the magnitude of the diffusion current that is related to concentration, and one must know which diffusion current to measure in the event that there are several current increases in a given polarogram. Figure 11.16 shows an example of one such polarogram.

Notice that the residual and background currents in Figure 11.16 are

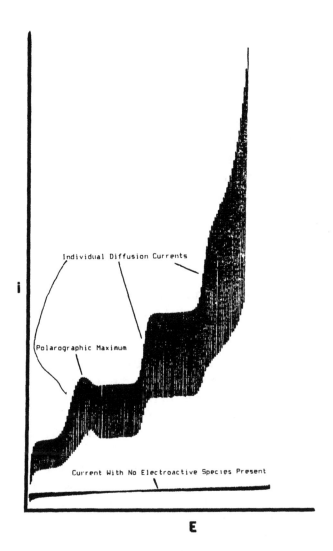

FIGURE 11.16   A polarogram showing several diffusion currents due to the presence
of several electroactive species in the solution. The analyst may identify
the analyte species' diffusion current by the magnitude of the associated
half-wave potential.

steadily increasing slightly as the potential is increased. This is due to the
fact that the mercury drop is becoming more and more charged as the
potential is increased. The instrument sees this as a slight current flow
and displays it on the recorder. This is a "non-Faradaic" type of current.

A "Faradaic" current is one that is due to the oxidation or reduction of a dissolved chemical species. It is due to an electron transfer process. A non-Faradaic current is one that is not due to an electron transfer process.

Polarographic "maxima" are sometimes encountered in the current/ potential display, and these can inhibit the accurate measurement of the diffusion current. Polarographic maxima are distortions of the polarogram manifested by a large "peak" of current immediately beyond the half-wave potential (Figure 11.16). The cause is not yet well understood, however, the problem can usually be solved by the addition of small amounts of gelatin or a surfactant, such as Triton X-100, to the solution.

11.2.3c Modern Polarographic Techniques

Modern polarography instruments are capable of several modifications of the classical procedure just described. These are "sampled dc polarography," "pulse polarography," and "differential pulse polarography." Each differs from classical polarography either because of the nature of the potential-time characteristic, or the mode of current measurement, or both. In most of these, the sensitivity is dramatically increased over the classical mode. In one of them, the resolution of two closely spaced half-wave potentials is improved. The techniques are summarized in terms of their potential-time characteristics and their current-potential recorder trace in Figure 11.17.

Sampled dc polarography is identical with the classical technique except that the current is only "sampled" and displayed just before the drop falls. The advantage is that it makes the diffusion current easier to measure — such a current is free of the continuous up and down pattern.

Pulse polarography is like sampled in that the current is sampled just before the drop falls. However, the potential-time characteristic is very different; a series of increasing potential pulses is applied to the DME, as shown, each time just before the drop falls. The current sampling is done while the pulse is being applied, which is for less than a tenth of a second. Thus, the following sequence of events occurs in rapid order: (1) the pulse is applied, (2) the current is measured and displayed, (3) the pulse drops back, and (4) the drop falls. After a delay (on the order of seconds), this sequence begins again, but with a higher pulse. Once the pulse is high enough to cross the half-wave potential, the current begins to rise due to the reduction or oxidation of the analyte. The

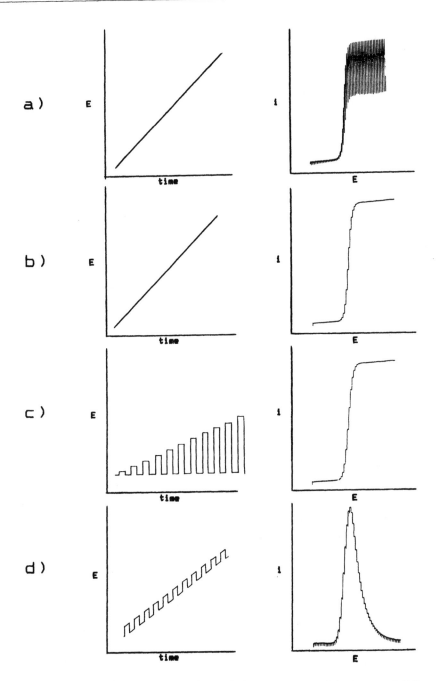

FIGURE 11.17   Modern polarography techniques compared in terms of E-time characteristic and i-E trace: (a) classical dc, (b) sampled dc, (c) pulse polarography, and (d) differential pulse polarography.

advantage is a much higher current for the same concentration, and thus a better sensitivity.

With differential pulse polarography, both the potential-time characteristic and the current sampling method are different from the others. The potential-time characteristic is one in which pulses of constant magnitude are superimposed on the ramp as shown. The current is sampled both before and during the application of the pulse. These two current values are subtracted from each other and divided by the potential difference, creating a signal which represents the rate of change or the "derivative" of the original polarogram. It is this signal that is then displayed on the recorder as the sampled currents were for the others. The trace resembles a peak because it is the rate of change of the original polarogram, and in the original, there is very little change before and after the rise in current, while a sharp change occurs where there is a rise. Thus, a peak occurs at or near $E_{1/2}$. The advantage of this technique is not only that it is very sensitive (typically more sensitive than any of the others), but also, two species that have similar $E_{1/2}$ values, while not distinguished before, can be with this technique.

11.3.2d Quantitative Analysis

In classical (and sampled dc) polarography, the diffusion current, the current increase that occurs when the half-wave potential of the analyte is reached, is directly and linearly related to the concentration of the species involved. Quantitative analysis is then based on the measurement of the diffusion current. The usual procedure is to prepare a series of standard solutions and measure the diffusion current for each. A plot of i_d vs concentration then is useful in determining the concentration in an unknown. In pulse polarography, the current increase observed for each of a series of standard solutions is plotted vs concentration. We have already stated that this current is typically much higher than that for classical and sampled dc, and thus quantitative analysis is much more sensitive. See Figure 11.18 for a comparison between sampled dc and pulse polarography.

The height of the peak in differential pulse polarography is proportional to concentration. Thus, in this technique, the peak height for each of a series of standard solutions is measured and plotted vs concentration.

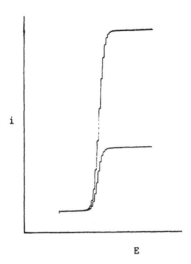

FIGURE 11.18   A comparison of a pulse polarogram with a sampled dc polarogram for the same concentration of electroactive species.

11.3.3 Voltammetry

11.3.3a Introduction

As stated previously, "voltammetry" is an amperometric technique in which the working electrode is some stationary electrode and not the DME. Typical electrodes here include small platinum (or gold) strips, wire coils, or beads sealed into the tip of a glass tube so that a small area or cross-section is exposed and polished. They may also consist of a small bead of graphitic carbon sealed into the tip of a glass rod and also finely polished as the others. This latter electrode is referred to as a "glassy carbon" electrode. A third type is the so-called hanging mercury drop electrode (the HME). This electrode is like the DME, except the mercury is not flowing but held stationary.

The nature of the voltammetry techniques we will now describe dictates that the electrode be stationary. In one case, the electrode serves as a collection point for the analyte, thus precluding the need to have a continuously renewed surface. In another case, the electrode monitors the chemical reactivity near the electrode surface, thus precluding the need to have the solution stirred.

11.3.3b Stripping Voltammetry

This technique is most useful for metals that are able to be electroplated onto the surface of a working electrode that has been polarized to some negative potential so that reduction of the metal ion to the metal atom will take place, i.e., at a potential more negative that the half-wave potential for such a reduction. This reduction occurs over a period of time, perhaps even on the order of 1 or 2 h, while the solution is stirred (with the use of a stirrer) to aid in the transport of the metal ion to the electrode surface. Over this period of time, a fairly large quantity of metal ions are reduced, and the metal either deposits onto the surface of the electrode or, in the case of the HME, a very common electrode for this technique, dissolves into the mercury drop. After the preset period of time has elapsed, a "reverse" scan is performed so that the applied potential crosses the $E_{1/2}$ "barrier" and all the metal is oxidized back to the metal ion, in effect "stripping" it from the surface. The resulting oxidation current (usually the *total* current measured in coulombs) is compared to that of a standard that has undergone the same experiment and the concentration determined. The potential-time characteristic and the resulting current pattern are shown in Figure 11.19. The advantage is that a solution of a *very low* concentration can be analyzed because the analyte is preconcentrated in or on the electrode over a period of time, making the stripping current easily measurable.

11.3.3c Cyclic Voltammetry

This is a technique that is seldom used for quantitative analysis, but is a popular technique for other reasons, such as the determination of electrode reactions mechanisms. The potential-time characteristic is shown in Figure 11.20.

As the potential is increased past $E_{1/2}$, the current is monitored continuously. The potential scan is then reversed and returned at an equal rate back to the starting point. The current/potential curve, if the electrode reaction is reversible, is shown in Figure 11.21a. If it is not reversible, that is, if the product of the electrode reaction has been reacted to form another species, the result is shown in Figure 11.21b. The oxidation current evident in (a) shows that the product of the reduction is still available at the electrode surface to be oxidized back. In (b) it is not available.

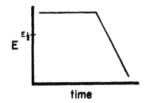

FIGURE 11.19   The potential and current characteristics of a stripping voltammetry experiment. (From Kenkel, J., *Analytical Chemistry for Technicians*, Lewis Publishers, Inc., Chelsea, MI, 1988. With permission.)

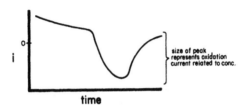

FIGURE 11.20   The potential-time characteristic in cycle voltammetry. (From Kenkel, J., *Analytical Chemistry for Technicians*, Lewis Publishers, Inc., Chelsea, MI, 1988. With permission.)

11.3.3d Amperometric Titration

The concept of current measurement in voltammetry can be applied to a titration experiment much like potential measurements were in Section 11.2 (potentiometric titration). Such an experiment is called an "amperometric titration"; a titration in which the end point is detected through the measurement of the current flowing at an electrode.

The potential of the measuring (working) electrode, which is typically a rotating platinum disk embedded in a Teflon sheath, is held constant at some value beyond the half-wave potential of the analyte. The solution

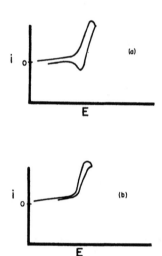

FIGURE 11.21   Cyclic voltammograms of (a) a reversible electrode process and (b) of an irreversible process. (From Kenkel, J., *Analytical Chemistry for Technicians*, Lewis Publishers, Inc., Chelsea, MI, 1988. With permission.)

is stirred due to the rotation of the electrode. The resulting current (the diffusion current) is then measured as the titrant is added. The titrant reacts with the electroactive species, removing it from the solution, and thus decreasing its concentration. The measured current therefore also decreases. When all of the analyte has reacted with the titrant, the decrease will stop and this signals the end point.

CHAPTER 12

AUTOMATION

12.1 INTRODUCTION

There is a continuing effort to streamline analytical laboratory procedures. Over the years, such procedures have developed from the time-consuming classical wet methods of analysis to the more sophisticated and faster instrumental methods. In turn, the instruments have undergone continuous improvement and streamlining to the point where large samples and standards are able to be tested in a very short period of time through what are called "automated" procedures. Today, automatic titrators, automatic analyzers which incorporate instrumental designs and concepts, and automatic samplers which are attached to traditional instruments, are commonplace. Even complex sample preparation schemes involving sample dissolution and extraction utilize robotics to help streamline the procedures. In this chapter, we discuss the design and theory of automatic titrators, the Technicon AutoAnalyzer,* flow injection analyzers (FIA), and robotics.

---

*   Registered trademark of Bran & Luebe, Inc., Buffalo Grove, Illinois.

12.2 AUTOMATIC TITRATIONS

When a large number of titrations is part of a laboratory's daily workload, it is possible for the laboratory to employ a partially automated or even a fully automated system. This system may be something simple, such as a unit that automatically refills the buret, or refills the buret with the use of a vacuum bulb, after a titration. In this case, a turn of the stopcock causes the buret to refill either by gravity or by vacuum from a large reservoir of titrant. It may also be an automatic dispensing buret with a digital display of titrant volume. Such a unit fits into a large reservoir of titrant solution and dispenses the solution into the titration flask. The operator starts and stops the titration according to his observations of the indicator, but the titrant volume is automatically measured and displayed, thus eliminating meniscus reading errors.

A fully automated system involves what is called an automatic titrator or autotitrator. Such a unit not only utilizes an automatic dispenser, such as described above, but also has automatic potentiometric, amperometric, or coulometric end point detection. The unit can even employ a computer controlled carousel in which samples rotate one after the other automatically through the unit. Such a system automatically activates reagent addition, stirring, measuring, titrating, aspirating, rinsing, and reconditioning of electrodes.

12.3 SEGMENTED FLOW METHODS

The Technicon AutoAnalyzer, developed in the late 1950s and early 1960s, represents the widely used analysis system that is based on "segmented" flow. It is an automation system in which the sample solutions, and various reagents with which the sample solutions need to be mixed, flow from their original containers through a maze of plastic tubes, the diameters of which define the rate of flow and thus the proportions to be mixed. At appropriate and strategic locations along the flow route, the reagents and samples come together and mix, creating the solution to be measured. In addition, air is drawn into the system also through one or more plastic tubes, and thus air bubbles are introduced as a result which help to divide the flowing liquid into segments, hence

FIGURE 12.1   An illustration of a plastic tube through which an air-segmented solution is flowing. (Adapted from Publication No.TN1–0169–00, Technicon Instrument Corporation, Tarrytown, NY; Bran & Luebbe Analyzing Technologies, Elmsford, NY. With permission.)

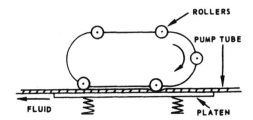

FIGURE 12.2   An illustration of a peristaltic proportioning pump. (Reprinted from Publication No.TN1–0169–00, Technicon Instrument Corporation, Tarrytown, NY; Bran & Luebbe Analyzing Technologies, Elmsford, NY. With permission.)

the "segmented flow" designation (see Figure 12.1). The complete system consists of a series of specific modular units assembled for a particular analysis. Modular units common to all analyses using this system are an automatic sampler (an "auto-sampler"), a peristaltic proportioning pump, a colorimeter, and a recorder. The proportioning pump is "peristaltic," which means the pumping action is performed through the repeated squeezing of the tubes through the action of metal rollers (see Figure 12.2). The system of tubes positioned in the pump along with the interconnecting tubes, mixing coils (see below), and glass and plastic pieces is called the "manifold." The auto sampler consists of a rotating carousel which positions vials containing the sample solutions in such a way that they are drawn out, one after another, through one of the tubes leading to the proportioning pump. The length of time during which the sample is drawn out is determined by an interchangeable cam in the carousel.

At the same time, the pump also delivers the air bubbles, a diluent if desired, and one or more reagents needed for color development such that all of these ultimately meet and move through a single tube. Thorough mixing then takes place by channeling this mixture, with the air bubbles, through coiled glass tubing for mixing. The air bubbles are of special assistance to the mixing process. Without air bubbles, a layer of a liquid adheres to the wall of the tubing causing contamination of the flow stream

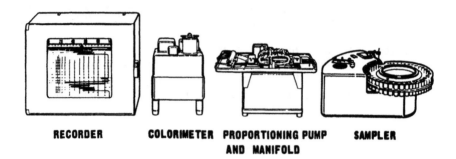

RECORDER   COLORIMETER PROPORTIONING PUMP   SAMPLER
AND MANIFOLD

FIGURE 12.3   A drawing of the hardware components of the Technicon
AutoAnalyzer system as described in the text. (Adapted from
Publication No.TN0–0169–10, Technicon Instrument Corporation,
Tarrytown, NY; Bran & Luebbe Analyzing Technologies, Elmsford,
NY. With permission.)

to follow. After mixing and color development, the air bubbles are drawn
off (via a "debubbler"), the solution flows through a detector, usually a
colorimeter equipped with a flow-through cuvette, and the transmittance
level is traced on the recorder and/or fed into a computer. The entire
system, assembled for an analysis such as just described, is illustrated in
Figure 12.3.

Specific methods and analytes require specific tubing diameters, junction
tubes, mixing coils, etc. To transform a system designed for phosphate
analysis, for example, into one for chloride analysis requires a complete
dismantling and reassembly of the manifold, unless the manifold systems
are stored so that they can be reinstalled later. Technicon manufactures
and sells all tubing and parts needed for a particular manifold. The tubing
is color-coded and tubing junctions and mixing coils are given part
numbers which are specified in AutoAnalyzer system flow diagrams
available from Technicon or found in method books, such as in *Standard
Methods for the Examination of Water and Wastewater*. Figure 12.4 gives two
such system flow diagrams, one for chloride in water and wastewater
(a) and one for phosphate in water and seawater (b). The color codes, such
as "BLK/BLK" and "ORN/GRN," symbolize the colors of plastic clips
that are found on the opposite ends of a given tube, and their associated
flow rates are given within the parentheses next to the stated color code
as shown. The clips are used to stretch and install the tubes in the pump.
Part numbers for the tubing junctions and mixing coils are also given in
these diagrams. You will notice, for example, in the chloride system
(Figure 12.4a) that a 14-turn mixing coil has part number 116-0152-02.

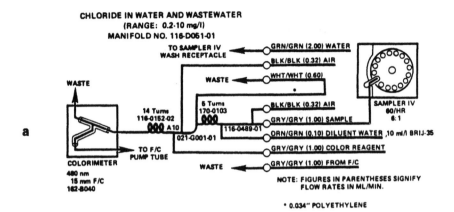

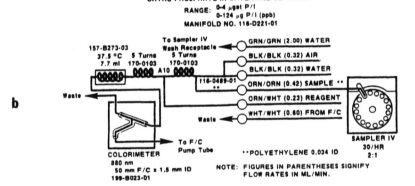

FIGURE 12.4   (a) The AutoAnalyzer system for chloride. (b) The AutoAnalyzer system for phosphate. (Courtesy of Technicon Instrument Corporation, Tarrytown, NY; Bran & Luebbe Analyzing Technologies, Elmsford, NY.)

You should also notice in Figure 12.4b that one of the mixing coils, the one positioned just before the colorimeter, requires an elevated temperature (37.5°C). A high temperature bath is another module that can be added to the system. Yet another module is a dialyzer which is needed for glucose analysis. The colorimeter (See Chapter 6) consists of a visible light source, glass filters for monochromators, a flow cell (cuvette), a photocell, and the associated electronics required to produce a percent transmittance (%T) signal to be sent to the recorder. The air bubbles must

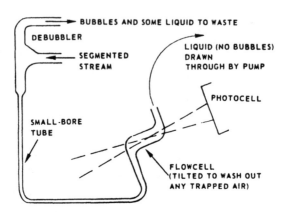

FIGURE 12.5   Flow cell and debubbler. See text for discussion. (Reprinted from Publication No.TN1–0169–00, Technicon Instrument Corporation, Tarrytown, NY; Bran & Luebbe Analyzing Technologies, Elmsford, NY. With permission.)

be removed prior to the solution entering the colorimeter. A "debubbler" is used for this. Figure 12.5 is a diagram of the debubbler in combination with the flow cell. The air bubbles float and pass upward and out while the solution is drawn down into the flow cell by the vacuum created due to a connection to the pump.

The recorder records the %T signal over time (it is a strip-chart recorder). It is calibrated in %T units. As sample after sample (separated by distilled water volumes) drawn from the auto sampler pass through the system, a recorder signal for each is generated, and these signals are then recorded one after the other on the chart paper. Figure 12.6 shows a sample of such a recorder trace. A popular modern alternative to the recorder is data acquisition with a computer.

A wealth of information, including additional details of the system setup and details of specific methods and manifolds for various analytes, is available from the Technicon Instrument Corporation.

12.4 FLOW INJECTION METHODS

More recently, an automated flow system in which the various liquid flow streams are *not* segmented has been developed. To say that they are not segmented means that air bubbles, which are characteristic of the Technicon system described in the previous section, are not introduced

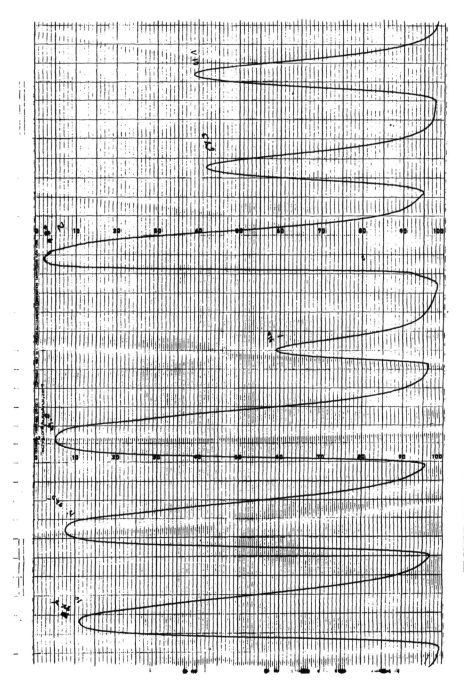

FIGURE 12.6   A typical AutoAnalyzer strip-chart recording of a series of solutions.

at any time, and thus the various flow streams are not characterized by segments separated by the bubbles. This nonsegmented type of system is known as Flow Injection Analysis or FIA.

In the segmented flow method, the air bubbles serve to help mix the various flow streams after they come together in the system. The obvious question with regard to FIA then is how do the various flow streams adequately mix prior to measurement. With FIA, the tubing diameters are much smaller (0.5 mm i.d.), and a much smaller amount of sample is "injected" with a valve similar to HPLC systems described in Chapter 10. Thus, the need for thorough mixing is greatly reduced, as adequate mixing of such small quantities occurs more readily without the need for air bubbles. A small "plug" of sample is thus carried through the flow system and sharp spikes are observed on the recorder trace. The advantages are (1) greater sample throughput, meaning a more rapid analysis of large numbers of samples per unit time; (2) shorter time from injection to detection; (3) faster start-up, shut-down, and changeover; and (4) less sophisticated equipment. For an excellent review of FIA, please refer to a paper which appeared in *Analytical Chemistry* in 1978[*] and again in Volume 2 of *Instrumentation in Analytical Chemistry* in 1982.[**]

12.5 ROBOTICS

Given the large volume of samples to be analyzed in many analytical laboratories, total automation of laboratory procedures has been a goal of analytical chemists and instrument manufacturers for years. With the advent of automation equipment, such as has been described in this chapter, and the ever more sophisticated computer and computer software, discussed in Chapter 5, this dream is coming closer to reality. It is now commonplace for sample solutions and standards to be rapidly diluted and mixed with appropriate reagents and measured one after another quickly as they pass through a flow channel. Along with this, the resulting data can be acquired, manipulated, statistically adjusted, and stored by a computer all with only pressing a button or a computer key in an operation that is not only automated, but may even be unattended for

---

[*]   Betteridge, D., "Flow Injection Analysis," *Analytical Chemistry*, 50(4), 832A (1978).
[**]   Borman, S.A., Ed., *Instrumentation in* Analytical *Chemistry*, Vol. 2, American Chemical Society, Washington, DC, 1982.

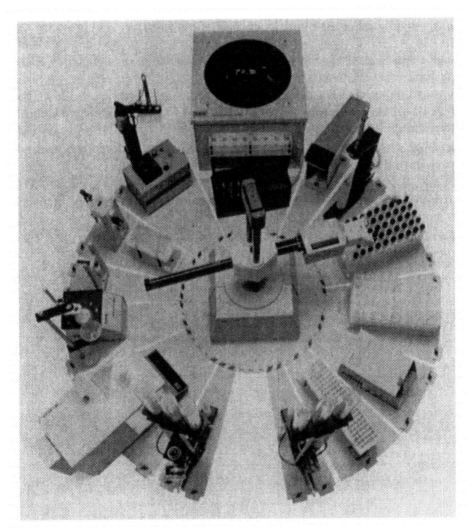

FIGURE 12.7   An example of a laboratory robotics system. (Courtesy of Zymark Corporation, Hopkinton, MA.)

a substantial period of time. All that remains is for the chemist to take the raw samples and perform the dissolution and/or extraction techniques described in Chapter 2 manually prior to executing the automated procedure.

The interesting news is that even these latter operations can be automated, since robots are available that will perform sample preparation tasks. Such robots are not androids or other two-legged variety that may immediately come to mind in light of some futuristic TV program or movie. A "robot" typically consists of a robotic "arm" that is capable of 360° rotation around a table equipped with the necessary auxiliaries, such as tube racks, balances, extraction and dissolution solvent reservoirs, vortex mixers, shakers, dispensers, and centrifuges (see Figure 12.7). The robotic arm is programmed to move samples from one treatment station to another in a sample preparation scheme designed to emulate the manual procedures. Besides having the ability to perform the routine preparation schemes without becoming bored, requiring a lunch or coffee break, or being limited to an 8-h day, the robot is able to work in hazardous environments, which is a significant advantage given the increasing publicity of health risks associated with handling the chemicals needed for such schemes. Analytical chemists can fully expect further developments in this area as time goes on.

CHAPTER 13

DATA HANDLING AND ERROR ANALYSIS

13.1 MODERN DATA HANDLING

13.1.1 Introduction

The molecular spectroscopic techniques of ultraviolet (UV), visible (vis), and infrared (IR) spectrometry; the atomic spectroscopic techniques of flame atomic absorption, graphite furnace atomic absorption, and inductively coupled plasma; the instrumental chromatography techniques of gas chromatography (GC) and high performance liquid chromatography (HPLC) — all of these and the myriad of others discussed in this book have become very popular analysis techniques of the modern analytical chemistry laboratory. The ability to handle the quantity of data produced and the ability to assure the quality of the decisions made as a result of the data produced are uppermost in the minds of laboratory managers everywhere. It is obviously important for these managers to be keenly aware of modern data handling procedures and to be able to execute them.

At the heart of modern laboratory data handling is the laboratory computer, often termed the laboratory "workstation" or "data station." The use of the computer in the laboratory was briefly discussed in Chapter 5. The modern laboratory utilizes computer workstations to (1) acquire data from instruments, (2) store the acquired data, (3) display and manipulate the acquired data, (4) control instruments, and finally to (5) gather and print the results in a way that is acceptable to clients. Professional analytical chemists must apply their knowledge of chemistry to the laboratory methods and procedures, but they must also be knowledgeable in computer usage for all the categories listed above.

13.1.2 Data Display and Analysis

In Chapter 5, the conversion of the analog signal typical of the output of many types of modern laboratory instruments to the digital signal required by a computer was discussed. The display of the digitized data on a monitor screen or a printer/plotter for subsequent interpretation and analysis is a common activity in these laboratories. Such a display amounts to a reconstruction of the analog data which allows the analyst to quickly determine maxima or minima in the data (such as may be required for chromatography peak heights or the absorbance at the wavelength of maximum absorption), perform simple mathematical operations on all or part of the data (such as multiplying by a dilution factor), average all or a portion of the data (which may be required for statistical analysis — see the next section), and many other similar manipulations. The computer especially enhances chromatographic data analysis, since it is capable of quickly determining retention times (and thus quickly identifying mixture components) as well as quantitation parameters, such as where to begin and end a peak integration, whether the peak threshold limit defined by the analyst has been reached, and the actual determination of the peak sizes by the integration (see Chapter 9).

Various companies market the hardware and software required for computer workstations. Such systems are readily available in various price ranges and are commonplace in modern laboratories.

13.1.3 Reporting and Managing Results

Modern computerized data stations are capable of automatically accumulating and assembling data files for the purpose of preparing and storing personal reports for the analyst, as well as customized reports for clients. Such reports can be assembled for presentation before a group or they can also be formatted for importing to standard microcomputer software, most notably word processing, data base, and spreadsheet software, which is useful if a formal written report of the results of an analysis is required.

In laboratories in which the volume of laboratory data generated is large, the computer is a godsend. The analyst does not even need to view individual routine results. Often the computer takes over, accumulates and analyzes the data, and generates reports which it automatically imports to data base files for a formal printed laboratory report. A visual check of the resulting report can pinpoint problems and suggest remedies.

13.1.4 The State of the Art

The strip-chart recorders and x-y recorders discussed in Chapter 5 are fast becoming obsolete, if they are not already. They are useful for immediate visual inspection of the analog data and are inexpensive, but they respond slowly and are not capable of automatic range switching and attenuation. Instrument output signals that "go off scale" or are too small to see need to be generated a second time.

Computing integrators (Chapter 9) are in common use in GC and HPLC laboratories. These perform retention time and peak integration determinations quickly and accurately, but usually have limited report-generating capability and cannot store and reconstruct chromatograms, items we discussed above as advantages of modern computer workstations. They are, however, relatively inexpensive and portable.

Thus, personal computers and computer workstations, as well as minicomputer and mainframe systems, have transformed laboratory data handling into job functions that are often as easy as pressing a key on a keyboard, especially given the increasing speed of the computers' central processors. Virtually all the functions listed in the preceding discussions can be performed quickly and accurately, thus making the handling of analytical data electronically automated. Modern chemical literature* addresses these issues and should be scanned regularly for modern developments.

13.2 INTRODUCTION TO ERROR ANALYSIS

" 'In laboratory testing, there are no magic black boxes where you simply follow directions and a number flashes on the screen that guarantees the truth,' says Dr. Benjamin Turner, a pathologist in private practice in Tallahassee, Fla. 'Laboratory tests can never be 100 percent perfect. Even the finest testing instruments can break down — and human beings aren't infallible either.' "**

How true Dr. Turner's words are! They dramatically point out perhaps the most important aspect of the job of the chemical analyst — to assure that the data and results that are reported is of the maximum possible quality. This means that the analyst must be able to recognize when the testing instrument is "breaking down" and when a "human" error is suspected. The analyst must be as confident as he/she can be that the "number that flashes on the screen" does in fact "guarantee the truth" as much as is humanly possible. The analyst must be familiar with error analysis schemes that have been developed and be able to use them to the point where confidence and quality is assured.

Errors in the analytical laboratory are basically of two types: "determinate," also called "systematic," and "indeterminate," also called "random." Determinate errors are avoidable blunders that were known to have occurred, or at least were determined later to have occurred, in the procedures. They arise from such avoidable sources as contamination, wrongly calibrated instruments, reagent impurities, instrumental malfunctions, poor sampling techniques, incomplete dissolution of sample, errors in calculations, etc. Sometimes correction factors can adjust for their occurrence, at other times the procedures must be repeated so as to avoid the error.

Indeterminate errors, on the other hand, are impossible to avoid. They

---

*   See, for example, a column entitled "The Data File" by G.I. Ouchi appearing regularly in the journal *GC·LC*, which is a journal dedicated to modern instrumental chromatography.

**   From Heller, L.J., "How Accurate Are Medical Tests?", *Parade Magazine*, February 3, 1991. Reprinted with permission from *Parade* 1991.

are random errors; errors which either were not known to have occurred, or which were known to have occurred, but could neither be accounted for directly nor avoided. Examples include problems inherent in the manner in which an instrument operates, errors inherent in reading a meniscus, errors inherent in sample and solution handling, etc. Since they are unavoidable and unknown, they are presumed to affect the result both positively and negatively and are dealt with by statistics. A given result can be rejected as being too inaccurate based on statistical analysis indicating too great a deviation from the established norm. The procedure to check for such errors involves running a series of identical tests on the same sample, using the same instrument or other piece of equipment, over and over. Those results that agree to within certain predetermined limits are averaged, and the average is then considered the correct answer. Any results that fall outside these predetermined limits are "rejected" and are not used to compute the average. The "rejectability" parameters are the standards which a given laboratory must determine and adopt for the particular situation. Some methods for the determination of these parameters will be discussed in this chapter.

13.3 TERMINOLOGY

In working with the statistical methods used to establish the "correct" answer in an analysis, a number of terms appear which need to be defined. These are as follows:

1. *Mean* — In the case in which a given measurement on a sample is repeated a number of times, the average of all measurements is an important number and is called the "mean." It is calculated by adding together the numerical values of all measurements and dividing this sum by the number of measurements.
2. *Median* — For this same series of identical measurements on a sample, the "middle" value is sometimes important and is called the median. Thus the median is always one of the actual measurements, while the mean may not be. However, if the total number of measurements is an even number, there is no single "middle" value. In this case, the median is the average of two "middle" values.
3. *Mode* — The value that occurs most frequently in the series is called the "mode." For a large number of identical measurements, the median and mode should be the same.

4.   *Deviation* — How much each measurement differs from the mean is an important number and is called the deviation. A deviation is associated with each measurement, and if a given deviation is large compared to others in a series of identical measurements, this may signal a potentially rejectable measurement which will be tested by the statistical methods. Mathematically, the deviation is calculated as follows:

$$d = |m - e|$$ (13.1)

in which d is the deviation, m is the mean, and e represents the individual experimental measurement. [The bars ($|\ |$) refer to "absolute value," which means the value of d is calculated without regard to sign; i.e., it is always a positive value.]

5.   *Relative Deviation* — Perhaps of more practical significance is the "relative deviation." This quantity relates the deviation to the value of the mean. If the mean is a relatively large number, a deviation that appears to be large may not be out of line because the value from which it shows deviation is large also. Thus a relative deviation is more useful. Mathematically, it is defined as follows:

$$d_R = \frac{d}{m}$$ (13.2)

in which d_R is the relative deviation. d_R can also be expressed as a percent, or part per thousand, etc.:

$$\text{relative \% deviation} = d_R \times 100$$ (13.3)

$$\text{relative ppt deviation} = d_R \times 1000$$ (13.4)

in which ppt represents "parts per thousand."

6.   *Average Deviation* — An overall picture of the quality of data (in terms of precision) is the "average deviation." It is the average of all deviations for all measurements:

$$d_a = \frac{(d_1 + d_2 + d_3 + ...)}{n}$$ (13.5)

in which d_a is the average deviation and n is the number of deviations (corresponding to the number of measurements) considered.

7. *Standard Deviation* — Of more popular use for demonstrating data quality is the "standard deviation," which is similar to the average deviation:

$$s = \sqrt{\frac{(d_1^{\,2} + d_2^{\,2} + d_3^{\,2} + \ldots)}{(n-1)}} \qquad (13.6)$$

The term $(n-1)$ is referred to as the number of "degrees of freedom," and s represents the standard deviation. The significance of both d_a and s is that the smaller they are numerically, the more precise the data, and thus presumably the more accurate the data. The standard deviation is used in most statistical methods in one way or another in evaluating reliability, that is, in establishing the level of confidence for arriving at the correct answer in the analysis.

8. *Relative Standard Deviation* — One final deviation parameter is the relative standard deviation, s_R. It is analogous to d_R in the sense that d_R is obtained by dividing d by the mean and multiplying by 100 or 1000. In this case s is divided by the mean and then multiplied by 100 or 1000:

$$s_R = \frac{s}{m} \qquad (13.7)$$

and

$$\text{relative \% standard deviation} = s_R \times 100 \qquad (13.8)$$

and

$$\text{relative ppt standard deviation} = s_R \times 1000 \qquad (13.9)$$

This parameter relates the standard deviation to the value of the mean and represents a practical and popular expression of data quality.

13.4 DISTRIBUTION OF RANDOM ERRORS

A graphical picture of the distribution of the results of an experiment, and thus a picture of the distribution of the random errors, is the "Normal Distribution Curve." It is a plot of the frequency of occurrence of a result on the y-axis vs the numerical value of the results on the x-axis. The normal (also called "Gaussian") distribution curve is bell-shaped, with the average of all results being the most frequently observed at the center of the bell shape and an equal drop-off in the frequency of occurrence in both directions away from the mean (see Figure 13.1). It is a picture of the precision (and thus presumably accuracy — see Chapter 1) of a given data set. The more points there are bunched around the mean and the sharper the drop-off away from the mean, the smaller the standard deviation, the more precise the data, and the more confidence one can have in any one result being correct. Typically, any measurement that is within predetermined confidence limits, and thus within the so-called confidence "interval," is deemed to be of sufficient accuracy to be reported or to be included in the calculation of the mean, which is then reported. Any measurement outside these confidence limits are of insufficient accuracy and rejected.

13.5 STUDENT'S t

A more formal method is a calculation taking into account the mean, the standard deviation, the number of measurements, and a "probability factor" (t). This latter parameter was the result of the work of W.S. Gossett, who published it under the pen name "Student" in 1908. It is often called "Student's t." The calculation establishes the confidence limits:

$$\text{analysis result} = m \pm \frac{t \times s}{n} \qquad (13.10)$$

and thus an analysis result, r, can be expressed as $r \pm c$, where c is the confidence limit or interval.

The values of t depend on the number of measurements, n, or, more

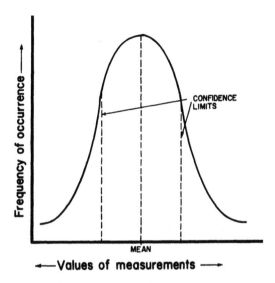

FIGURE 13.1   A normal distribution curve showing confidence limits a particular number of standard deviations from the mean, such as 1 or 2.

Table 13.1   *t* Values for Calculating Confidence Limits

n – 1	t (at 90% level)	t (at 95% level)	t (at 99% level)
1	6.3	12.7	63.7
2	2.9	4.3	9.9
3	2.35	3.2	5.8
4	2.13	2.78	4.6
5	2.02	2.57	4.03
6	1.94	2.45	3.71
7	1.90	2.37	3.50
8	1.86	2.31	3.36
9	1.83	2.26	3.25
10	1.81	2.23	3.17
00	1.64	1.96	2.58

levels and for a number of different degrees of freedom. The smaller the number of measurements, the less confidence the analyst has in his/her analysis result, as indicated by the larger values for t in Table 13.1. The larger the value of *t*, the larger the confidence interval. The more measurements that are made the better. From a practical point of view, it is the analyst's decision as to how much time is spent making measurements weighed against a desired confidence in the results. An analyst can reject results that push the interval past a desired value.

13.6 STATISTICS IN QUALITY CONTROL

In addition to providing the basis for rejection of "bad" laboratory data, statistics can also be used as a basis for detecting unacceptable batches of raw materials or problems in a quality control laboratory. If a given analysis yields results that are quite different from what is expected, the "rejection" indicated by statistics results in the rejection of the batch or sample from being used or shipped, or whatever.

Confidence limits that are exceeded can also sometimes be indicated in an instrument reading that exceeds a statistically predetermined limit, rather than a final answer. An example is the GC analysis of cellophane extracts for formaldehyde. Injection of the extracts one after the other into a GC may give a series of formaldehyde peaks as shown in Figure 13.2. Sample number 5 in this chromatogram shows a peak that exceeds the preset confidence limit. Under these circumstances, the batch of cellophane from which this sample was acquired is discarded.

In addition, "quality control charts" are sometimes kept in a lab in order to detect problems or trends in analysis procedures or material analyzed. The idea is that over a period of time, certain standard deviation values or confidence limit values are established which create a range of acceptable analysis results. A measurement falling outside the range or a trend away from the mean would then indicate a problem — possibly a determinate error, or a series of bad raw material batches, etc. An example of a quality control chart is shown in Figure 13.3.

13.7 ASSISTANCE

For a detailed discussion of these and other concepts relating to quality assurance in an analytical laboratory, you are referred to a book written by John Keenan Taylor.[*] Dr. Taylor's book is written to provide assistance for beginning a quality assurance program in a laboratory and provides the guidance needed to develop the reliability reputation that all laboratories seek. The underlying goal sought by all analytical laboratories is that all analytical results *must be reliable*, and there must be undisputable evidence to prove it. Dr. Taylor's book is based on this premise.

---

[*] Taylor, J.K., *Quality Assurance of Chemical Measurements*, Lewis Publishers, Inc., Chelsea, MI, 1987.

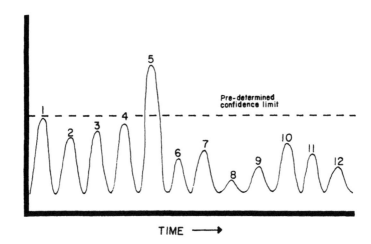

FIGURE 13.2   A hypothetical gas chromatogram of a series of cellophane extracts. See text for a brief description.

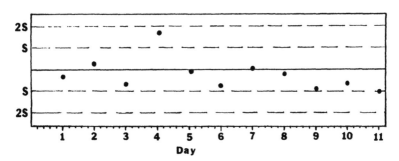

FIGURE 13.3   A typical quality control chart. Notice the measurements outside the range (defined by S) on day 4 and the trend away from the mean after about day 7.

INDEX

HPLC-MS, 165
Mass spectrometry, 162–165,
   282, *see also* Gas chroma-
   tography-mass spectrom-
   etry (GC-MS); Liquid
   chromatography-mass
   spectrometry
   instruments for, design of,
      163–164
   mass spectra and, 164–165
Material Safety Data Sheet
   (MSDS), 45
Matrix matching, 95, 185
McReynolds Constants, 236
Mean, defined, 333
Measurement, 6–9
Measuring pipets, 28–30
Median, defined, 333
Membrane potential, pH
   electrode and, 296
Methanol, in extraction, 41
Methylene chloride, in extrac-
   tion, 40
Mode, defined, 333
Mohr pipet, 28–30
Molar absorptivity, Beer's Law
   and, 127
Molar extinction coefficient,
   Beer's Law and, 127
Molarity, 54–56
   of acids, 49, 50
   of bases, 50
Molecular absorption spectrum,
   113
   double-beam spectrophotom-
      eter and, 120
   qualitative analysis and, 123
Molecular spectroscopy, 103–
   165, *see also* Fluorometry;
   Spectrometry; Spectrophotometry
   light and
   absorption and emission of,
      108–111
   nature and parameters of,

103–107
NMR, 155–162, *see also*
   Nuclear magnetic reso-
   nance (NMR) spectros-
   copy
techniques and instruments
   used in, 111–112
Monochromatic light, 112
Monochromator
   flame atomic absorption, 182–
      183
   flame photometry, 175–176
   UV/vis spectrophotometry,
      114–116
MSDS (Material Safety Data
   Sheet), 45
Multielement hollow cathode
   lamp, 180
Multiple-beam balances, 20, 21
Murexide, in water hardness
   analysis, 79

Neat liquids, IR spectrometry
   and, 131
Nebulizer, in atomic absorption, 170
Nernst Equation, 288–291
Nitric acid
   density of, 55
   molarity of, 50
   percent composition of, 55
   for sample preparation, 38
Nitrogen/phosphorus detector
   (NPD), 246
NMR spectroscopy, *see* Nuclear
   magnetic resonance
   spectroscopy
Non-Faradaic current, 311, 312
Normal distribution curve, 335–
   337
Normality, 56–62
   equivalent and, 57–61
   solution preparation and, 61–
      62